RICHARD MABEY

Richard Mabey is the father of modern nature writing in the UK. Since 1972 he has written some forty influential books, including the prize-winning *Nature Cure*, *Gilbert White: A Biography*, and *Flora Britannica*. He is a Fellow of the Royal Society of Literature and Vice President of the Open Spaces Society. He spent the first half of his life amongst the Chiltern beechwoods, and now lives in Norfolk in a house surrounded by ash trees.

ALSO BY RICHARD MABEY

RICHARD MABEY

Turning the Boat for Home

A Life Writing About Nature

VINTAGE

1 3 5 7 9 10 8 6 4 2

Vintage is part of the Penguin Random House group of companies
whose addresses can be found at global.penguinrandomhouse.com

Penguin
Random House
UK

Copyright © Richard Mabey 2019

Richard Mabey has asserted his right to be identified as
the author of this Work in accordance with the Copyright,
Designs and Patents Act 1988

First published in Vintage in 2021
First published in hardback by Chatto & Windus in 2019

penguin.co.uk/vintage

A CIP catalogue record for this book is available from the
British Library

ISBN 9781529111958

Illustrations on pp. 1, 65, 137 and 215 by Clare Roberts, originally
published in *The Journals of Gilbert White* (Century, 1986–9)

Printed and bound in Great Britain by Clays Ltd, Elcograf S.p.A.

The authorised representative in the EEA is Penguin Random House
Ireland, Morrison Chambers, 32 Nassau Street, Dublin D02 YH68.

Penguin Random House is committed to a sustainable future for our
business, our readers and our planet. This book is made from Forest
Stewardship Council® certified paper.

For Peter Newmark, old friend, who was there for a lot of this

Contents

CONTENTS

Prologue

It is the most intense and prolonged physical scrutiny I have ever had from a friend. For hours on end. Jon's face is no more than eighteen inches from mine. He's peering at the bags under my eyes, noting the nervous twitches of my mouth, logging moods made flesh. Every few seconds his gaze moves to the ball of clay on a rotating stand just to his right, and a few more pellets, the saggy increments of age, are added to what is beginning to resemble my chin.

I was immeasurably touched when Jon Edgar invited me to have my head sculpted as one part of an 'environmental triptych' he was making. The other two subjects were James Lovelock, creator of the Gaia hypothesis, and the philosopher Mary Midgley, so I was in flattering company. Now, perched on a stool (also rotatable), set between my study and our garden pond, I wasn't so sure. I felt exposed and vulnerable and very self-conscious.

But the feeling didn't last. Over what proved to be two days of sitting I began to slip into a state of Zen-like relaxation. Jon's method of working was to begin with a central wood peg, and gradually build up the head with small pieces of clay kneaded into place with his thumb, an oddly calming action to watch, as if he was massaging me instead. I had no clear notion of how the work was progressing, as Jon kept my features facing away from me, though halfway through he gave me a glimpse of what then, shockingly, resembled a photo of my father at half my age. Was this what I was turning into? But I soon stopped worrying, particularly about how easily I was distracted – by stomach rumbles, by a flock of long-tailed tits tumbling into the hawthorn bush just to my right, by the ceaseless fidgets of my facial muscles that always made me such an awkward subject for photographers. To have an impartial observer contemplate your face – the window to who you are – so closely, without making a value judgement was both novel and therapeutic. I started to feel comfortable about the way I looked, who I was. The solid veracity of the clay helped: this is you, like it or not.

In the end I did like it. The finished head, cast in bronze, wasn't that of the slightly haunted thirty-year-old I sometimes fantasised I still resembled, but a senior citizen with a thickening neck and a touch of gravitas. Ten years on it sits in the library, topped with a rakish straw hat I picked up in a Provençal market, now a reminder of a moment in my past, but – and this is Jon's art – with intimations of what went before and of future ageing caught in the metal sinews.

Jon subsequently cast a second bronze, and planted it outdoors in his garden, having discovered that rainwater and urine contributed greatly to the patina. 'The Lady Scott and Mabey bronzes,' he told me, 'were shot-blasted clean and then put on copper pipe spikes in the vegetable patch. These were weed on regularly, and now, several years later, have some of the finest patinas – akin to the qualities of the bronzes which have been retrieved from under the sea after thousands of years.'

I rather like the idea of an image of myself being assembled out of daubs of earth, and being the result of a kind of dialogue, not of earnest and solitary introspection. Over the years, I've written many fragments of memoir, about landscapes explored, fantasies made real, about roots and routes. But I've never felt that I was up to writing a full-blown autobiography. Partly it's a failure of memory, and of confidence. I am hopeless at recalling long-lost inner feelings, the details of family transactions, even recent conversations.

I am in awe of those writers who seem able to recount, exactly, sitting on their father's knee (it is always the father) aged four, the smell of his tobacco and the words on the page of the book open on his lap. And I lack the novelist's skill to make it all up. But, to be honest, the real reason is that I have no burning passion to explore those aspects of life that are the meat of conventional autobiography.

But when it comes to writing, the substance and business of it, that's another matter. Writing has been the best part of my life, and I continue to be fascinated by the process as much as the content. This is especially true of writing about the natural world, my 'vocation' I suppose, because it raises so many challenges: finding empathy with non-human organisms; respecting their individuality and otherness while acknowledging we are all connected parts of the same biosphere; reconciling rhapsodic joy at what is there with the full knowledge that its existence is critically threatened. So when I felt the need, in my seventies, to take some sort of long view over my life, it was the story of the work that I wanted to tell. As it happened, the job was already more than half done. Over the past two decades I have done a good deal of occasional writing – broadcast essays, introductions to other writers' work, seasonal journals – which contain broadly autobiographical material. Often they revisit episodes of exploration, or moments when ideas or books were born, and reassess them in terms of my current thoughts and beliefs. *Turning the Boat for Home* is a collection of these pieces, arranged so that a sketchy reflection on a life's work does emerge. I value this short-order work as being every bit as important as the books, and the place where vital, first impressions are thought through and set down. I'm suspicious of grand, overarching narratives. Nature does not work in broad strokes, nor, I confess, does my mind. I'm a magpie and much taken with Keats's ideas of 'negative capability' and creative uncertainty.

The pieces contain straight nature writing, too, and I hope the balance is fair. I sometimes chastise myself for spending more time with my nose in a book than out sniffing for foxes or compiling the parish Flora I have always promised myself; the declining energy of old age and an increasing sensitivity to bad weather are both partly to blame. But I don't want to over-apologise, as the two experiences, of language and of nature itself, don't feel that separate to me. For as far back as I can remember I've been a compulsive maker of stories-in-the-head, of an internal commentary about what is happening to me. I play with ethereal half-sentences and risky metaphors while my feet are still struggling through the mud. It feels so instinctive that I've persuaded myself that it's the human (or at least my) equivalent to the way animals endlessly read the scent marks of other organisms in their territory. Putting this wordy potpourri down on paper is just an extension of the process. My friend Tim Dee makes an incomparable defence of the relation between the indoor and outdoor experiences in his book *Four Fields*: 'I learned that I needed the indoor world to make the outdoors be something more than simply everything I wasn't. I saw it was true that indoor talk helped the outdoor world come alive and could of itself be something living and lovely, too. Words about birds made birds live as more than words.'

For convenience I've arranged the pieces around four themes, and more loosely in a sequence which echoes my own development as a writer. The first is about my early writing influences and my commitment to a materialistic view of nature; the second is about the curious business of writing about plants; the third is concerned with politics, and especially the politics of woodland; and the fourth on 'new nature writing'.

There is inevitably an underlying note of loss and anger running through the book, from shock at how the profusion of life that Richard Jefferies wrote about a century and a half ago has all but disappeared, to the continued poisoning of barn owls by agricultural rodenticides. It is impossible to write about nature in the twenty-first century free of the corrosive shadows of climate change and mass extinction. Indeed, some of my colleagues believe it isn't any longer appropriate – maybe not even ethical – to write about nature in a celebratory tone. I respect their views, but a bit of my heart is with the words Jeff Goldblum famously mumbled in *Jurassic Park*: 'Life will find a way.' Crises don't extinguish the redemptive power of nature, and we need to revel in that not just for its own sake, but because it may yet help us out of the abyss.

PART ONE

The Shock of the Real

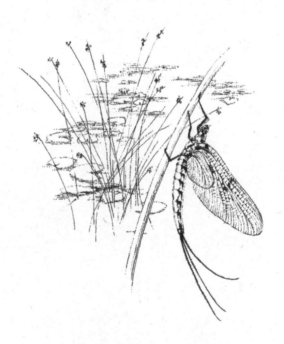

In 2005 Andrew Motion recorded a series for Radio 4 entitled *A Poetry Map of Britain*. One of the programmes was on 'Flatlands' and, having moved to Norfolk ('Very flat' as Noël Coward famously decreed) not many years before, I was invited to contribute some thoughts about what was now my home landscape. We stood among the alders and swaying reedbeds of Redgrave Fen, and I explained what I felt about the character of East Anglia's wetlands. I have never understood the conventional view that they are melancholy, desolate, forlorn, possessed of a 'sad splendour' as the Suffolk poet George Crabbe put it. To me they have always been electrifyingly alive, full of nuance and surprise. Water energises them, imbues them with a sense of possibility. It also helps create a system of intricate and constantly shifting particularities. Close by us on the fen were the haunts of the great raft spider, a species at that time confined to the small flooded peat pits that are a special feature of this place.

As we were coming to the end of the recording Andrew asked me whether I had a spiritual connection to these landscapes. I was taken by surprise. No one had put such a question to me before. As someone who doesn't believe in spirits I've never been precisely sure what 'spirituality' meant. So I blurted, 'No, I'm a transcendental materialist.' It was a contrived answer (and thankfully didn't survive the producer's edit) and I should probably have had the guts to say I was a materialist pure and simple, for all the unpleasant associations of that word. But that wasn't quite what I meant. Although I have an absolute belief in the physical world, in the glorious diversity of stuff, it doesn't stop there. I often find the particularities of the natural world so strikingly present that they seem (if this isn't taking liberties with words) to transcend themselves, not into anything 'beyond', or mystically underlying, but into an intense clarity of being. I believe this is a widespread feeling, but one that is hard to put into words. Which is perhaps why writers have invented new terms for this impression: 'thisness', the quality that makes one thing itself and not something else; Gerard Manley Hopkins's 'inscape'. The American writer Thomas Berry coined 'inscendence' by which he meant not going beyond the evidence of one's senses but digging down deeply into it.

There are maybe 30 million different species on the earth, if we count all the fungi and bacteria which are still being discovered. Every single one has invented itself and every one is perfect for its role and connected to all the others. We understand the broad way in which evolution works, but only a little so far about the detailed development of every species' special characteristics and life-ways. But the mysterious is not the mystical. I find that scientific revelations about how life works enhance my sense of wonder, not diminish it. And I am baffled (affronted, sometimes) when life's marvellous complication is reduced to some inner or 'spiritual' essence; or when writers think it smart, and somehow a badge of the depth of their feelings, to mock the importance of identity in the natural world and suggest that the ability to tell a blackcap's song from a blackbird's is demeaning. The astonishing billion-year-long arc of diversifying life is trivialised to a green blur. 'World is crazier and more of it than we think,' wrote Louis MacNeice, 'Incorrigibly plural.' I once delivered a secular grace at a wedding breakfast by reading the final paragraph of Darwin's *Origin of Species* and its heart-stirring words about 'endless forms most beautiful and wonderful.'

This section is about my first influences as a writer, and about how they fostered an attention to the detail of the world. Richard Jefferies was the first nature writer I read, and I found his sense of particularity, and of 'instress' as a sacrament, echoed later by Ronald Blythe and the American biologist Lewis Thomas. But the writer who has been a constant presence through much of my writing life has been Gilbert White, author of *The Natural History of Selborne* (1789), the earliest text of 'nature writing'. I first became properly acquainted with White when I was asked to write an introduction to *Selborne* for a Penguin Classics edition. I was struck by what a complex man he was, an Oxford-educated intellectual marooned in a remote Hampshire parish, and not at all the simple rustic priest he'd so often been portrayed as. I went on to write his biography, which won the 1986 Whitbread Biography Prize. Every year I glimpse new nuances in his work, as diarist, Romantic scientist, miniaturist, story-teller. For this collection I've chosen something a little different, an outline of the highlights of his life, written as a rough storyboard for a film I made with the historian Michael Wood.

The section begins with an introduction I was invited to write to a new edition of Peter Matthiessen's *The Snow Leopard*, which was written just after the author had been introduced to Buddhism, and was struggling with the tensions between science and spirituality, reality and illusion. Engaging with Matthiessen's interior debate proved to be a better way of clarifying (or not) my own thoughts on these central themes than trying to express them in isolation.

Snowcat

First published as an introduction to The Snow Leopard,
Vintage Classics edition, 2010

Just ten days into his odyssey to the Crystal Mountain in Nepal, Peter Matthiessen finds his physical composure already in tatters. 'My knees and feet and back are sore, and all my gear is wet. I wear my last dry socks upside down so that the hole in the heel sits on the top of my foot; these underpants, ripped, must be worn backwards; my broken glasses' frame is taped.' Like a novice before initiation he decides to strip off the rest of his worldly accoutrements. He has his hair cropped short to the skull. He junks a heavy wristband, worn 'latterly as an affectation'. Lastly goes his watch, 'as the time it tells is losing all significance'. What aren't jettisoned, thankfully, are his notebook and his gift for language. At this acute moment of spiritual preparation, of waving aside physical bonds, what goes down into words are finite and thoroughly mundane details. Holey socks. Reversed underpants. The professorial lenses held together by tape. If this is Matthiessen's world-turned-upside-down, its material underbelly remains mischievously alive in his words.

But he's laying down a warning marker for the reader here. This isn't going to be an easy trip. Don't expect the glittering landscapes and mythic creatures of the Himalayas to become accessible by proxy, or a short walk. Don't hope that enlightenment comes with an easy ousting of the physical. Burdened with Western baggage we may have to strip down too, ditch the assumption that a quest has to have an end, for instance, and that the 'I' of the storyteller is necessarily the narrator. Most crucially we will have to learn to live in this edgy territory where the material and the numinous stake out equal claims to our attention.

What was Matthiessen seeking in this journey? *The Snow Leopard* is the first of his great travel journals to be as much a spiritual quest (whatever that may mean) as an exploration of an exterior wilderness, and it's significant that he wrote it when he did. His matter-of-fact explanation is that in 1972 the zoologist George Schaller (an old friend from the Serengeti wanderings recounted in *The Tree Where Man Was Born*, 1968) invited him to join an expedition to north-west Nepal to study the bharal, the Himalayan blue sheep. The bharal appealed to both men's sense of the aboriginal: to the biologist Schaller as the possible ancestor of all domesticated sheep and goats; to the romantic Matthiessen as a kind of ghost from Arcadian prehistory.

But there was a more personal reason behind Matthiessen's decision. Earlier that year his wife Deborah Love, who had introduced him to Zen Buddhism, had died of cancer. He was still grieving, but, by his own spiritual standards, grieving too much. Buddhism teaches the renouncement of craving and neediness, and that the transition between life and death is just a single shift in the myriad states of universal energy. But Matthiessen hadn't found that good karma yet, hadn't reached such a state of unearthly detachment. Maybe up in the purifying snows of Buddhism's heartland, he might find acceptance, and join the world here, now, as it is.

And then there was the small matter of the snow leopard, whose terrible beauty is the very stuff of human longing. Its uncompromising yellow eyes gaze out from the covers of innumerable editions of Matthiessen's narrative. Schaller had mentioned that they had a good chance of seeing this near-mythical creature on the Crystal Mountain, and Matthiessen felt the prospect 'was reason enough for the entire journey'. What was he up to, pinning 'this journey of the heart' on such a remote but oh-so-fleshly animal? He hints at the kind of grail the leopard represents: 'It has pale frosty eyes and coat of pale misty grey, with black rosettes that are clouded by the depth of the rich fur ... it is wary and elusive to a magical degree and so well camouflaged in the place it chooses to lie that one can stare straight at it and fail to see it. Even those who know the mountains rarely take it by surprise: most sightings have been made by hunters lying still near a wild herd when a snow leopard happened to be stalking.' You truly see by not trying to see.

The snow leopard is not some kind of facile symbol; Matthiessen is too good a writer to play that card. It's a flesh-and-blood animal that he is desperate to glimpse, but also desperate to allow to remain invisible, secretly itself, untouched by his cravings. It is the tension between these two seemingly irreconcilable quests – the inquisitive naturalist, hungry for new experience, and the Buddhist novice, eager to reach beyond the superficialities of experience – that gives the book its charge. No wonder its appeal endures. In the 1970s, it was one of the texts in the backpacks of Westerners on the hippy trail. Later it helped inspire a new breed of field naturalists and film-makers to venture with humility into extreme environments. Today it still raises challenges. We have a broken relationship with the natural world, and do not know how to heal it. Is our peculiarly human breed of consciousness – and our overblown sense of 'self' especially – a barrier to re-engagement, or the only medium through which we can hope to make any sense of the outside world? Matthiessen's book, poised between a search for what might be called a common consciousness and an intensely personal, idiosyncratic documentary, is a fundamental handbook for this task.

It's from this confusion that I start my own expedition into *The Snow Leopard*, feeling as ill-equipped for the task as the author did for his. I have never entirely understood what 'spiritual' means to those who don't believe in spirits. If it is just a pious synonym for feelings or emotions then perhaps it would be better said so in plain sight. I'm content to have no faith, unless an exultation in the 'endless forms' of creation is itself a kind of belief. I like the way my sense of who I am is taped together by the ceaseless scribblings in my head, and have no wish to meditate them away. Yet I've been attracted by Buddhism myself, by its godlessness, its concentration on immediate experience – and then nonplussed when it dismisses such experience as worldly illusion. But I'm reassured by recalling one reason for Matthiessen's Nepalese journey that he doesn't mention. He's a professional writer. This is what he does. He goes on adventures in order to tell us stories about them.

The adventure itself is arduous, but never catastrophic. By Matthiessen's standards it's also short, and he's back in New York in little more than two months. But from the outset it seems a lonely journey. The party is twenty strong, including four sherpas and fourteen porters. But Matthiessen strides clear of them immediately. Perhaps this is the custom on these long treks, but you sense he wants to be alone with his thoughts, or at least with his perceptions. 'Hibiscus, frangipani, bougainvillea: seen under snow peaks, these tropical blossoms become the flowers of heroic landscapes.' Everything is a kind of bloom, an extrusion of the land: 'The fire-coloured dragonflies in the early autumn air, the bent backs in bright reds and yellows, the gleam on the black cattle and wheat stubble, the fresh green of the paddies and the sparkling river – over everything lies an immortal light, like transparent silver.' A few days later he spots a common sandpiper – 'it teeters and flits from boulder to black boulder, bound for warm mud margins to the south'. He'd seen this 'jaunty' bird in many other places, from Galway to New Guinea, and is 'cheered a little when I meet it again here'.

And we're cheered too by this little tug at the heartstrings, this brief human admission of nostalgia. Because Matthiessen has already begun the more earnest business of interrogating his reactions and motivations according to Buddhist precepts. 'The Universe itself is the scripture of Zen,' he writes, 'for which religion is no more and no less than the apprehension of the infinite in every moment.' How does this flash of affective sandpiper memory weigh in Zen's scales? As a moment of weakness, drawing Matthiessen away from the intense experience of the here and now? Or as a moment of *prajna* – a vision of one's identity with universal life, past, present and future?

I'm in need of some spiritual porters at these points. I understand the words, though not always their meaning. I admire their implicit morality. But the heavy abstraction and convolutions of Buddhist theory seem to drag down the brilliance and immediacy of Matthiessen's encounters with the physical world. Is the goal of enlightenment to see and love this world exactly as it is, or to see through it, to something beyond?

Matthiessen understands this problem and experiences it himself. This is where George Schaller helps. GS is Sancho Panza to Matthiessen's Don Quixote. He grounds the book. He's practical, shrewd, sometimes a little geekish. He is awesomely fit, but makes his feet bleed by walking too ferociously. He loses patience with the porters, and wants to throw one off a cliff for endlessly intoning OM MANI PADME HUM. He hates the slowness of the expedition, and yearns for the 'crisp air' in the high mountains of the blue sheep and snow leopard. 'Once the data start coming in,' he says, 'I don't care about much else; I feel I'm justifying my existence.' It sounds like a scientist's version of the Zen belief in the primacy of 'Is-ness'.

And so they proceed, the intense encounters with this immense and compelling landscape counterpointed by reflections on whether it is just a mirage. By late October they are at 13,000 feet. 'Three snow pigeons pass overhead, white wings cracking in the frozen air. To the east, a peak of Dhaulagiri shimmers in a halo of sun rays, and now the sun itself bursts forth, incandescent in a sky without cloud, an ultimate blue that south over India is pale and warm, and cold deep dark in the north over Tibet – a blue bluer than blue, transparent, ringing.' It is a world that seems more real than real, yet Matthiessen tells us of the teaching that existence is but a dream, a void, that all phenomena, including time and space 'are mere crystallisations of mind'. It's a metaphysical position that sits oddly with his moments of existential excitement: 'It is! It exists! All that is or was or will be ever is right here in this moment. Now!'

And that exclamation suddenly reminds me of other writers. Matthiessen makes much of the similarity between Buddhist teachings and quantum physics, modern theories of consciousness, even Christianity. St Catherine's 'All the way to Heaven is Heaven', he insists, 'is the very breath of Zen'. But those eyeball-to-eyeball encounters with the here and now are also the stuff of Matthiessen's own literary tradition. In 1855 the American poet Walt Whitman had a secular vision of the unity of the world, also glimpsed from a high place. In *Leaves of Grass* he wrote:

> I have no chair, nor church nor philosophy;
> I lead no man to a dinner table or library or exchange,
> But each man and each woman of you I lead upon a knoll,
> My left hand hooks you round the waist,
> My right hand points to landscapes,
> And a plain public road.

Just seven years earlier Henry Thoreau had made his epic climb of Mount Ktaadn and gave a statement about the absolute authority of the physical. It prefigures Matthiessen even in its highly stressed punctuation: 'Talk of mysteries! Think of our life in nature – daily to be shown matter, to come into contact with it – rocks, trees, wind on our cheeks! The *solid* earth! The *actual* world! The *common sense! Contact! Contact!*' Later, in the famous passage in *Walden* where he discusses measuring the depth of his pond, he seems to set his face against any notion of a world beyond the material: 'The greatest depth was exactly one hundred and two feet; to which may be added the five feet which it has risen since, making one hundred and seven. This is a remarkable depth for so small an area; yet not one inch of it can be spared by the imagination ... While men believe in the infinite some ponds will be thought bottomless.' The imagination, he is suggesting, needs finitude, not abstraction, for its full flowering.

So does Peter Matthiessen in the descriptive passages of *The Snow Leopard*. The book's concluding sections, played out in the leopard's own domain, are majestic. At last he gets to see the blue sheep at close quarters, and the data coming in seems to justify his existence just as it does George Schaller's. 'I lie belly down, out of the wind, and the whole warm mountain, breathing as I breathe, seems to take me in ... The lead female comes out of the hollow not ten yards up the hill, moving a little way eastward. Suddenly, she gets my scent and turns quickly to stare at my still form in the dust below. She does not move but simply, stands up, eyes round. In her tension, the black marks on her legs are fairly shivering; she is superb.' In a genre that is often stupefied with an excess of metaphor it is striking how little Matthiessen relies on such devices. (And when he does they are so exactly chosen that they reverberate like an arrow hitting a target. The Himalaya is 'as convoluted as a brain'; a lammergeier is 'a nine-foot blade sweeping down out of the north'.) Perhaps he thinks that words themselves are already metaphors enough. But his language is direct, an unadorned outgrowth of what he is describing and an embodiment of its significance. Again I'm reminded of Thoreau, and his fantasy of what 'natural' writing would be like: 'he would be a poet who could impress the winds and streams into his service, to speak for him; who nailed words to their primitive senses ... who derived his words as often as he used them, – transplanted them to his page with the earth adhering to their roots; whose words were so true and fresh and natural that they would appear to expand like buds at the approach of spring'.

Language is suspect in Buddhism, one of the layers of illusion to be transcended by meditation. It is sometimes referred to as a 'dirty pane of glass' between person and world. 'The weary self of masks and screens, defences, preconceptions, and opinions,' Matthiessen writes, 'that, propped up by ideas and words, imagines itself to be some sort of entity ...' So as a materialist I am selfishly glad that Matthiessen the writer wins out over Matthiessen the spiritual explorer, that ideas and words are set down as exact and substantial entities, and that he has left us with a luminous account of one of the most beautiful regions of the earth, and not rejected his words as mere ciphers for a larger reality. We are a language-using species. Language is how we understand ourselves and relate to our fellows, and used with Matthiessen's power is one of the ways we can find our way back to a kinship with the rest of creation. Maybe the central paradox of the book is that the author's meditative disciplines, his search for a purity of understanding beyond words, has led him to a supreme clarity of language.

But when it comes to the snow leopard, every philosophical position wins. The animal is never seen, of course. There are teasing hints and signs. One bright morning on the Crystal Mountain 'the leopard prints are fresh as petals on the trail'. Two days later, a leopard makes a scrape right across the author's bootprint. He takes it as a sign that he is not to leave. 'It is wonderful,' he writes, 'how the presence of this creature draws the whole landscape to a point.' But it fails to materialise, and in the end Matthiessen makes peace with his neediness: 'Have you seen the snow leopard?' 'No, isn't that wonderful!'

Materialist and Buddhist alike can take comfort in this rejection of hubris, in its acceptance of the world-as-it-is, with its own priorities, and not always the world as we want it to be. 'Expect nothing,' a Zen priest had warned Matthiessen before he set out. In the end, wrung out by anticlimax and the effects of sudden altitude loss, he has a small revelation, more cheering than any that came from his deliberate searching: 'I begin to smile, infused with a sense of my own foolishness, with an acceptance of the failures of this journey as well as of its wonders. I know that this transcendence will be fleeting, but while it lasts, I spring along the path as if set free; so light do I feel that I might be back in the celestial snows.'

Richard Jefferies

First published as an introduction to Wild Life in a Southern County,
Little Toller Editions, 2011

Richard Jefferies' *Wild Life in a Southern County* (1879) was my first encounter
with what even in the nineteenth century was called 'nature writing' (Jefferies
used the phrase himself and may even have coined it). I must have been
about twelve years old, and happy enough with my ragbag collection of
I-Spy books and *Observer* guides, texts about what was what, and where to
find it. When I discovered and read my elder sister Pat's copy of *Wild Life*,
it made me feel a little light-headed. Inside were meditations on how animals
might think, and how landscapes made you feel. There were chapters on
'Rooks Returning to Roost' and 'Snake Lore'. I'd forgotten most of its
contents within a month, but the title stuck in my imagination like the aura
of a half-recollected dream, or a mantra: wild-life-in-a-southern-county. I
was living in Hertfordshire at the time, and the only 'southern' place I'd
been to was the beach at Worthing. But my emotional compass was already
set in that direction. Whenever I walked out from home I went south, up to
the top of our road to the beginnings of the Chiltern chalk hills, rising
towards the summer sun. South meant a view down a long wooded valley,

my private heartland, and a thin stream that wound its silver way towards the scarp. South was the direction I went to meet the summer migrants flying back home. On May Day I would cycle across the fields to cricket holding my blazer collar for luck to charm in the first swifts. In those adolescent years – already an incipient romantic – I would stand at a ritually precise spot at the top of the hill and gaze down that valley in a state of muddled rapture. What I was looking at seemed both wild, somehow beyond reach and understanding, but also profoundly and anciently anchored. It made the back of my legs tense, as if I was trying to stop myself from falling.

No wonder Jefferies chimed with me. *Wild Life* (and note the powerful separation of those two normally conjoined words) is a collection of free-range essays exploring the author's unresolved feelings about the relations between the natural and human worlds. It's set in the very human context of the Wiltshire smallholding where he grew up, but is peopled mostly by non-human species. Jefferies makes the dialectic between these two worlds explicit in his short preface: 'There is a frontier line to civilisation in this country yet, and not far outside its great centres we come quickly even now to the borderland of nature ... If we go a few hours' journey only, and then step just beyond the highway – where the steam-ploughing engine has left the mark of its wide wheels on the dust – and glance into the hedgerow, the copse, or stream, there are nature's children as unrestrained in their wild, free life as they were in the veritable backwoods of primitive England.'

But the frontier is porous, fluid, debatable. Living things – humans included – pass across it, in both directions. So do ways of perceiving them, which can be coloured both by the civilised, rational mind and the feral imagination. Later in the preface Jefferies writes that 'nature is not cut and dried to hand, nor easily classified, each subject shading gradually into another. In studying the ways, for instance, of so common a bird as the starling it cannot be separated from the farmhouse in the thatch of which it often breeds, the rooks with whom it associates, or the friendly sheep upon whose backs it sometimes rides'. This 'shading' of subjects is an exact description of Jefferies' meandering prose style, and there is, I suspect, an element of rationalisation here of its sometimes chaotic discursiveness. But he was an intuitive ecologist, and this insistence on the connectivity of the natural world is a theme that runs through the book, and justifies his grouping of its contents by habitats. Except that they are not truly natural habitats, but the human niches around Coate Farm, near Swindon, where Jefferies was born and lived until he was eighteen: orchard, woodpile, home-field, ash copse, rabbit warren.

But there is sleight of place, and memory, here. Although *Wild Life* is set in Wiltshire, in the present tense, it was written in Surbiton in 1878–9, more than a decade after the encounters it chronicles. Jefferies was then thirty years old and had moved to suburban Surrey to be closer to the London newspaper world. The book is quarried from articles he contributed to the *Pall Mall Gazette*, early examples of what has become an enduring form in British journalism – the 'Country Diary'.

Jefferies' distance from the scenes he is describing (it was also, as we'll see, a social distancing) helps account for his fascination with borderlands, and for what is a dominant motif in the book. If the pathway is the thread which runs through Edward Thomas's analogous, echoic *The South Country* (1909 – Thomas published a biography of Jefferies in the same year), the route he followed to walk himself up and out of his black moods, the keynote of Jefferies' *Southern County* is the hedgerow, in which he can burrow down, away from the messiness of human society. The hedge is the 'frontier line to civilisation'. It is a mark of division in the affairs of humans, but a connective tissue for wildlife. It represents refuge, but also a kind of linear commonland. Jefferies recalls, in a later book, how his father used to point with disgust to 'our Dick poking about in them hedges', and like the poet John Clare, he is most at home – and at his best as a writer – in the hedge-bottom looking out, not on the hilltop taking imperious (or queasily spiritual) views of the landscape below. In one passage (subtitled 'A Hill apparently Enlarged') he recounts how, peering through a gap in a hedge, he once experienced a kind of optical illusion, in which a hill he knew suddenly appeared vastly higher than it had before. A cloud was resting on its top, and for a while had taken on the exact shape and tone of the hill. With the rest of the range obscured by the hedge, this glimpse through the gap revealed something closer to an alp. The aesthete always lurked, sometimes enlighteningly, sometimes subversively, inside the watchful naturalist.

More literally, hedges were the 'highways' of Jefferies' wild neighbours. Birds and animals passed up and down them between the copses and the farm. One major 'caravan route ... abuts on the orchard [and] the finches, after spending a little time in the apple and damson trees, fly over the wall and road to [a] second hedge, and follow it down for nearly half a mile to a little enclosed meadow, which, like the orchard, is a specially favourite resort'. It isn't hard to imagine 'our Dick' himself dogging the waves of tits and charms of blossom-haunting goldfinches ('a flood of sunshine falling through a roof of rosy pink') flowing down the hedgelines, building up a map of the birds' movements and 'resorts', and in the process conjuring the outlines of what sometimes seemed to him the skeleton or ghostly relics of the Wiltshire wildwood.

Jefferies' attention to what he saw is rapt, exact, almost painterly. The look of nature seemed to him as good a guide as any to its meaning and order. He notices the sparkle of ice on the high branches of beech trees in winter, and suggests that this 'proves that water is often present in the atmosphere in large quantities'. His vivid description of a magpie's movements perfectly catches the bird's character, but also begins to explain what it may be up to in its seemingly erratic foragings: 'he walks now to the right a couple of yards, now to the left in a quick zigzag, so working across the field towards you; then with a long rush he makes a lengthy traverse at the top of his speed, turns and darts away again at right angles, and presently up goes his tail and he throws his head down with a jerk of the whole body as if he would thrust his beak deep into the earth'. He devotes almost two pages to the ripening colours of wheat, noting a moment when it briefly pales during a breeze, 'because the under part of the ear is shown and part of the stalk'. He listens to the heavy buzz of hornets, and peers at them intently enough to know they are the most inoffensive of insects. And he watches a thrush smashing a snail on a sarsen stone: 'about two such blows break the shell, and he then coolly chips the fragments off as you might from an egg'.

There is a kind of hedge-scientist at work behind these observations, thinking by analogy, forging explanations by the application of reason (or at least a particular kind of reason) to acute observation. Jefferies rarely attempts to test his theories methodically, and never quotes the opinions or experiences of any other naturalists. He is an intellectual hermit. This occasionally leads him towards conclusions that today would be regarded as anthropomorphic, or plain fanciful. He observes the large clutch size and sociable behaviour of long-tailed tits (cousins and unpaired birds often help with the feeding of the young), and concludes that several female birds lay their eggs in one communal nest. A cuckoo lingering close to the nest where she's laid her egg makes him 'doubt the cuckoo's alleged total indifference to her young'. He is also sceptical about cuckoos' host species failing to recognise that the monstrous chick growing in their nest is not one of their own. 'The robin is much too intelligent. Why, then, does he feed the intruder? There is something here approaching to the sentiment of humanity, as we should call it, towards the fellow-creature.'

What lies behind these convictions is Jefferies' unusual attitude towards the idea of 'instinct'. He regards this as an inadequate explanation of the behaviour of wild creatures because they so often make mistakes. He tells the story of a party of sand martins attempting to quarry their nest holes in the mortar of a thick stone wall at Coate Farm. It was a fruitless task, and

'At last, convinced of the impossibility of penetrating the mortar, which was much harder beneath the surface, they went away in a body ... Instinct, infallible instinct, certainly would not direct these birds to such an unsuitable spot ... The incident was clearly an experiment, and when they found it unsuccessful, they desisted.' A modern scientific explanation would be that it was precisely the martins' instinct for exploring soft stone that led them to the wall, which was then overridden by their experience. But Jefferies' sympathetic interpretation was in a way correct, and far-sighted; he had simply adopted an overdeterministic view of the nature of instinct, seeing it as infallible, or 'blind'. Intelligent experimentation and exploration are now regarded as entirely compatible with broad instinctual drives.

Jefferies' belief in the free will of other beings, in the maternal cuckoo and the compassionate robin, extended to an insistence that animals felt joy in their lives: 'You may see it in every motion: in the lissom bound of the hare, the playful leap of the rabbit, the song that the lark and the finch *must* sing.' But, inside *Wild Life* at least, his sympathy with other creatures is patchy and inconsistent. There is a detachment in his prose, which displays plenty of intense curiosity, but little revelation about his own feelings. After a spellbinding and affectionate account of the family life of kingfishers, he gives, quite casually, as if he had forgotten what he said about joy, instructions about the best way to shoot the birds, especially the youngsters.

The fact is that, at this stage in his life, Jefferies had not yet worked out which side of that 'frontier line' he was on – anchored with civilisation, or on the wing with unrestrained nature. *Wild life in a Southern County*, his second non-fiction book, is a transitional work, marking the beginnings of a shift away from such simplistic separations of the world. Contrary to the popular image of him as a deep-rooted countryman, Jefferies was a displaced person almost from birth. Aged four, he was dispatched from his family's declining farm to live with an aunt in Sydenham. When he was nine he returned home, only to be shunted off to a succession of private schools in Swindon. No wonder he developed into a moody and solitary adolescent. He began reading Rabelais, and spent long days roaming the hill country round Marlborough. When he was sixteen he ran away from home with his cousin, first to France and then to Liverpool, where he was found by the police and shipped back to Wiltshire. When the family smallholding was badly hit by cattle plague in 1865, he left school for good, and started work in Swindon on a new Conservative paper, the *North Wilts Herald*, where he was a jack-of-all-trades reporter and resident short-story writer. At the end of the 1860s he became vaguely ill, left the paper and took a long recuperative holiday in Brussels. He was extravagantly delighted by the women, the fashions, the sophisticated manners, and from letters to his aunt it is clear what he was beginning to think of the philistinism of Wiltshire society.

He returned to Coate Farm in 1871, with no job and no money. His life began to slip into a mould more typical of the anxious, hand-to-mouth existence of the urban freelance than of a supposed 'son of the soil'. He wrote a play, a memoir of a prospective Member of Parliament, a right-wing pamphlet that ridiculed the advance of popular education. His breakthrough came with a letter to *The Times* in a similar Conservative vein, scorning the habits, intelligence and apathy of the Wiltshire farm labourer. The letter won him sympathy from landowners, and offers of more journalistic work, and for the next few years he wrote copiously on rural affairs for *Fraser's Magazine* and the *Live Stock Journal*.

His journalistic commitments sparked the move to Surbiton, and to writing regular essays for the *Pall Mall Gazette*, in which he reminisces – albeit in an idealistic way – about life back at Coate. His first fully-fledged non-fiction work, *The Gamekeeper at Home*, is made up of pieces written for the *Gazette* between late 1877 and spring '78. It is essentially a tribute to 'the master's man' and an account of the practical business of policing a sporting estate, and maintains the Conservative, deferential tone of his early journalism.

The pieces for *Wild Life* appeared in the *Gazette* between 1878 and '79, and though there is a new sympathy with the farm worker in it, and the first glimpses of his nature writing potential, flashes of the old right-wing shooting man continue to appear.

Jefferies had only eight years left to live at this point. He developed tuberculosis in 1881, and pain and disenchantment colour the rest of his work. He seems at last to understand the preciousness of life, to be engaged with it, not just as a curious observer but as a fellow being. His beliefs shift radically towards a kind of pantheism, and politically towards libertarian socialism. In his late essays be begins to write of the history, politics, ecology and aesthetics of the land as part of a single complex experience.

These final essays, such as 'Hours of Spring' (1886) and 'Walks in the Wheat-fields' (1887), are his most mature and powerful. But *Wild Life in a Southern County* contains their first buds. To read these essays today is chastening. There is, in the best of them, an electric attentiveness, a noticing, that is hard to aspire to. They are chastening, too, in what they are able to describe – an abundance of bird and insect life that, despite the contemporary passion for slaughter (in which the author played his part), is unimaginable in the modern industrial countryside. The great set piece of *Wild Life*, 'Rooks Returning to Roost', is like an epic Victorian narrative painting, full of intense images: the sound of thousands of black wings 'beating the air with slow steady stroke can hardly be compared to anything else in its weird oppressiveness'; full too of a sense of the deep history, the natural 'tradition' of these great nightly migrations. And of one stunning statistic: the 'aerial army's line of march extends over quite five miles in one unbroken corps'. Jefferies did not know this, but he was sending, in a faltering new language, a message in a bottle from a disappearing country.

A Very British Nature

First published in the Guardian, *2009*

On May Day 1945, just a week before the VE Day celebrations, the director
of Kew Gardens gave a talk on the wild flowers of London's bomb sites.
The Times, sensing the event's eccentricity, or perhaps some cryptic metaphor,
covered it as their lead news story. Professor Edward Salisbury, speaking in
the fragile shell of the Savoy Chapel Royal ('itself hit 4 times and damaged
11 times'), had described how a whole new suite of plants had taken advantage of the city's open wounds. Their names were evocative: many-seeded
goosefoot, gallant soldiers, *Senecio squalidus*. For the most part they were
interlopers and opportunists, immigrants from Southern Europe and North
America. Their seeds had drifted in on returning soldiers' boots and transatlantic food parcels, or had sprung to life after being buried for generations
under now shattered concrete. Canadian fleabane had reputedly arrived in
Europe in the stuffing of a parrot sent to Germany. Salisbury's catalogue of
phoenixes and fifth columnists suggested links between the affairs of humans
and wild nature that seemed to have, for want of a better word, meaning.
The odd thing, from a perspective sixty years on, is that neither he nor *The
Times* openly considered such a perspective. There is no attempt to milk the

green upsurge for its irony or questionable taste. It isn't presented as a symbol of the new cosmopolitanism or the resilience of life in the face of adversity. Nor, for that matter, as an allegory of wartime's reawakening of the dark powers of the wilderness. This was the ambivalent view of the novelist Rose Macaulay as she too rummaged through the catacombs of the City, where 'fireweed [rosebay willowherb] ran over Inigo Jones's court-room', a sign of 'the irremediable barbarism that comes up from the depth of the earth'. No, for most people in the 1940s, nature was either a hobby or just an ingredient of that nebulous idea of 'heritage', part of the national landscape.

As if to confirm this quotidian view, Salisbury's full bomb-site tally – 157 species – was published a few months later as an appendix to one of 1945's surprise bestsellers, *London's Natural History*, a lead title in a mould-breaking new series from William Collins publishers. 'The New Naturalist' volumes were, in the editors' words, intended to 'recapture the enquiring spirit of the old naturalists', and to foster 'the natural pride of the British public in their native fauna and flora'. They seemed to capture a yearning in at least part of the public, and the first two titles sold more than 30,000 copies each in their first eighteen months. Over the decades that followed the series became a publishing phenomenon, with a sixty-year-long list of quality volumes which probably has no rival in any other area of specialist publishing. They turned schoolboy birds-nesters into ornithologists, and line the shelves of every serious British naturalist over the age of thirty. Many have been standard texts on degree courses and have helped shape the outlook of the professionals who run Britain's conservation business. The series is going through one of its active spells at the present (no. 107, on British grouse, is just out) and the republication of the first ten titles in facsimile (a brave move at £75 a copy, obviously aimed at the collectors' market) provides an opportunity to assess its influence at a time when a rather different kind of writing about nature is in vogue. The massive changes in Britain's wildlife and landscape over half a century are obvious from its pages. But so, obstinately, is a peculiarly British attitude towards natural history – meticulous, enthusiastic but emotionally disengaged.

With hindsight the New Naturalist's (NN) early success is puzzling. The end of a war which had ravaged the whole fabric of European civilisation seems an odd moment to be bothered about pride in the nation's wildlife. But Britain's landscape (and to some extent its 'native fauna and flora') had been a central motif in wartime propaganda, and in the vision of those who were planning the post-war world. Frank Newbould's iconic Ministry of Information poster, 'Your Britain. Fight for it Now', showed (to the disgruntlement, it must be said, of the millions whose personal Britain was a blitzed city) a rhapsodic view of a shepherd leading his flock home over the South Downs. In the darkest days of the war in 1940, the popular radio naturalist James Fisher, later to be a key figure in the series, published a bestselling book on watching birds. 'Some people,' he argued, 'might consider an apology necessary for a book about birds at a time when Britain is fighting for its own and many other lives. I make no such apology. Birds are part of the heritage we are fighting for.' A similar spirit enthused the various committees optimistically set up by the government as early as 1942 to plan Britain's post-war estate of nature reserves and national parks. The historic White Paper of 1947, 'The Conservation of Nature in England and Wales', contained a startlingly bold and elegant encapsulation of their vision. The National Parks Commission was to 'construct for the people a lasting and pleasurable resort'. Nature, in short, was to become an ingredient of the post-war social settlement, the welfare state out-of-doors. But it wasn't to be utopian philosophers or the people themselves who were to set the agenda for this Arcadia. Instead it was to be delegated to a 'national Biological Service', staffed by meritocrats from the burgeoning sciences of genetics and ecology.

The chairman of the committee was Julian Huxley, the latest scion of a starry family, who went on to become director general of UNESCO. In the spring of 1942, Huxley was approached by Billy Collins, head of the family firm, to solicit his support for the natural history books. An impromptu first editorial meeting soon followed at the Jardin des Gourmets in Soho, with James Fisher and the Austrian refugee printing specialist Wolfgang Foges also present. The editorial board and the credo of the series were established over lunch. As with the proposed National Parks, a rigorously scientific approach, tempered by populism, was the goal. The geographer Dudley Stamp, who soon joined the board, reflected breezily in a draft of one of his own New Naturalist titles (*Nature Conservation in Britain*, 1969) that 'A bright new series [of books] was a godsend to harassed seekers after presents as well to a public hungry for peace and forgetfulness of war. What better than natural history?' But the credo the editors printed on the back of every new title set a more earnest house style. The public pride they set such store by was to be maintained by a 'high standard of accuracy combined with clarity of exposition in presenting the results of modern scientific research'. The books were to be grouped into two rough categories, studies of species and groups of organisms, and accounts of habitats and geographical regions. Fisher hoped that above all they would be infused by the excitement of first-hand observation in the field. A view of natural history as a way of understanding humans' role in life's larger scheme was some way in the future, but the series was at least shaping up to become, in effect, the guidebooks to the British people's 'lasting and pleasurable resort'.

The first two titles express the diversity of subject and author the editors hoped for. *Butterflies* is an uncompromisingly technical view of these animals by the Reader in Genetics at Oxford, E. B. Ford; *London's Natural History* is a brisker and more accessible documentary by an industrious amateur naturalist and freelance journalist, Richard Fitter. Yet, despite the editors' wish to capture a new public, both books follow a conventional, even old-fashioned structure. Fitter devotes the entire first half of his book to the historical growth of London. Ford launches in with a 'History of Collecting'.

Nothing expresses the difference in sensibilities between the 1940s and today better than his account of a mass assault on what is now one of Britain's rarest butterflies: 'I once captured numbers of the Black Hairstreak by climbing into a tree and, armed with a net on a long pole, sweeping them off the leaves on a neighbouring oak which was sticky with [honeydew]. I may add that very few of the specimens were worth keeping, so quickly does this insect damage itself on the wing.'

But both books are to some extent redeemed by their documentary detail. Ford's genetics chapters are a struggle for the lay reader, but he is fascinating on how variation occurs, on how butterflies communicate, on the Purple Emperor's taste for carrion. Fitter's main theme is the extraordinary adaptability of wildlife to the seemingly hostile conditions of a big city. He recounts how vast numbers of zebra mussels had colonised the water mains of south London, and how house mice were thriving in the stores of the Royal Albert Dock, growing thicker fur and dining off bags of frozen kidneys.

Graphic and telling anecdotes of this kind swarm in the first ten NN volumes. The ichneumon fly lays its eggs in the living body of a caterpillar, which the hatched larvae then proceed to eat alive, from the inside – a life cycle that shocked Charles Darwin, but not A. D. Imms, author of *Insect Natural History* (1947). The parasitic dodder, entirely rootless, slithers across the land in search of its host like a snake (*British Plant Life*, 1948, by W. B. Turrill). A puffball was discovered under on oak tree in Kent during the war, so big that it was assumed to be a secret German weapon (*Mushrooms and Toadstools*, 1953, by John Ramsbottom, Keeper of Mycology at the Natural History Museum). The stories are never gratuitous, and they proved magnets for the general reader. (I bought *Mushrooms and Toadstools* – my first NN volume – precisely to mine it for Gothic tales like the above.) But as you read through the volumes, the anecdotes begin to stand out with a vivacity often absent from the more central scientific sections. The biological skin of Britain was being laid out and anatomised with minute assiduousness and the very latest understanding, but it felt more like an autopsy than an adventure in field natural history. The British wildwood was demythologised, the old folk tale of the woodcock carrying its young about in flight confirmed. An entire Cheshire parish is shown pickled in the nineteenth century in A.W. Boyd's exquisitely strange *A Country Parish* (1951). All was well and known. A very British enterprise in certainty and dispassion was under way.

At the very start of *Wild Flowers* (1954) we are promised something different. John Gilmour and Max Walters, from the Cambridge University Botanic Garden and Herbarium, quote Shelley's 'Spirit of Beauty' to explain the motivation of the field botanist: 'News of birds and blossoming, – / Sudden, thy shadow fell on me; / I shrieked, and clasped my hands in ecstasy!' But no passion or beauteous spirit casts its ecstatic shadow on their text. It has a *plein-air* feel, and the denizens of wood and down are thoroughly and pleasantly catalogued. But of the experience of discovering them in their native habitats, of their presence, there is nothing. One of the icons of *London's Natural History* is the black redstart, which colonised London's bomb sites in the 1940s alongside Salisbury's rosebays. Fitter charts its invasion from rocky natural habitats in Europe, and shows that, contrary to the popular myth, it had begun nesting here before the Blitz. Its favoured sites have the incantatory power of, in more than one sense, a concrete poem: the Palace of Engineering in Wembley, the gates of Westminster School, the redoubts of the Inner Temple. But this thrilling bird with its flame-red tail and song like a rattled metal gate never appears as a living presence. It is a cipher, the staple that holds together an eco-geographical theory. A reader might be forgiven for wondering if Fitter had ever seen one himself, and it's clear that for many NN authors, venturing into the first person was seen as either embarrassing or compromising.

More British etiquette emerges in the treatment of conservation. Ritual lip service is paid to the idea, but usually in the recommendation of better nature reserves. What need was there of anything else? The war was over, and the period of white-hot agricultural and forestry intensification – what a contemporary NN writer, Oliver Rackham, has called the 'locust years' (*Woodlands*, 2006) – was ten years in the future. The countryside was back in the benign care of farmers and foresters, and it was still culturally unacceptable (especially in the wake of wartime food and timber crises) to question their immemorial role as the Good Stewards of nature. In reality, as we now know, Britain's wild places were already going down the pan. Heaths, downland, ancient woods and hedges were continuing to be devastated in the long programme of agricultural 'improvement' that had intensified during the war. Occasionally, and probably unwittingly, the NN editors connived in this traditional deference, and chose authors with what today would be called 'declarable interests'. For their first woodland title, *Trees, Woods and Man* (1956), they commissioned H. L. Edlin, a Forestry Commission officer, who devoted his volume largely to conifer plantations, and gave short shrift to the value of native deciduous woods. Kenneth Mellanby, who was head of entomology at the Rothamsted horticultural experimental station, was commissioned to write both *Pesticides and Pollution* (1967) and *Farming and Wildlife* (1981). His defence of DDT came five years after the publication of Rachel Carson's *Silent Spring*, and of unequivocal proof of the pesticide's calamitous long-term effects on both wildlife and humans.

To be fair to the editors and authors, few of them had the benefit of the kind of nationwide survey data that's taken for granted today. Yet one – Dudley Stamp, advisory editor and author – did have such access. His *Britain's Structure and Scenery* (1946) is an engaging book, a highly readable account of the fundamental geology of Britain, and still a good corrective to the fashionable view that all landscape is 'man-made'. But his view of superficial 'scenery' – something which most people would regard as influenced considerably by human ornamentation – is by contrast strikingly static. Which is curious, as before the war Stamp had organised tens of thousands of schoolchildren to prepare maps for the Land Utilisation Survey, the first field-by-field study of the condition and use of Britain's land since the Tithe Maps of the mid nineteenth century. No mention of that study appears in his book. Throughout the early NN titles there is a sense of an unspoken portal, through which certain kinds of evidence, because of their origins or apparent subjectivity, are not allowed to penetrate.

This is not to wish the NN to be something entirely different, or to join the kind of largely autobiographical, lyrically tinged work now known as 'nature writing'. It is to ask whether the series really fulfilled its stated purpose of making the enquiring spirit of field biology available to a wide public. In *Butterflies*, Ford gives an account of the bizarre life history of the rare Large Blue butterfly, probably the first time it had been in print for a popular audience. After its third moult, the Blue's caterpillar begins to develop a honey gland (called Newcomer's organ) in its abdomen, and begins crawling around until it meets a foraging ant. The meeting evokes an instant response. The ant begins caressing the honey gland, and drinking the secretion it produces. Other ants gather, and join in. After an hour or so, the caterpillar rears up into a grotesque ritualistic posture, at which signal the first ant seizes it in its jaws and carries it back to the nest. There it remains in total darkness from August until the following June, being cared for by the ants as if it were one of their own. It feeds on ant larvae, and in return becomes the honey cow for the colony. And all the while (though Ford did not know this in 1945), it is quietly singing to the ants, mimicking the noises of their grubs. The story is so extraordinary, so counter-intuitive, that Ford's detachment is exasperating. How was this partnership discovered? Had Ford witnessed it for himself? Was it observed in a laboratory, or in the field? And if so how? Was an entomologist prone in front of a West Country anthill, making sense of what he was seeing through agricultural metaphors? The story is offered to us not as a detailed account of an experience of nature, but *ex-cathedra*, finished, unmysterious, unarguable. It would be cynical to see this as a case of 'trust me, I'm a scientist', but it is a long way from being an authentic act of witness. The evidence, as with the bulk of the documentary material included in the NNs, has been already selected and processed, and the first-hand encounters with the living organisms that lay behind the revelations are drained of the vitality that makes them meaningful to non-specialists. Many of the NN authors, in fact, were reporting the end results of their own work, but you would be hard put to realise this. It is odd that modern British naturalists are so uncomfortable about describing their own experiences, given that an earlier generation – especially Gilbert White and Charles Darwin, often cited by NN authors as mentors – were masters at making narratives of the untidy and uncertain excitements of their fieldwork.

Perhaps it's no surprise that the two most exciting and readable volumes in the first tranche of reissues come from the margins of the naturalists' inner circle. *Natural History in the Highlands and Islands* (1947) was by Frank Fraser Darling, pacifist, socialist, and director of the West Highland Survey. His strong roots in the Highlands crofting economy give the book a real steeliness, and an understanding of the relations between wildlife and people. He expresses trenchant views about conservation, and in his concluding paragraphs rejects utilitarian rationalisations, including those cited by most other NN authors: 'Wildlife does not exist for man's delectation. Man may find it beautiful, edifying, amusing, useful and all the rest of it, but that is not why it is there, nor is that a good enough reason for our allowing it to remain. Let us give beast and bird and flower the place to live in its own right ...' Darling went on to give the historic Reith Lectures 'Wilderness and Plenty' in 1969, which marked the beginnings of 'green consciousness' in Britain. *British Game* (1946) is the oddest of the bunch, being concerned with a contentious and decidedly non-scientific class of creature, and written by the editor of *The Field*, Brian Vesey-Fitzgerald. But if you can swallow his tirade against hares as vermin, and the tallies of horrific battues (sixty-nine capercaillies in one day in 1910), it is also the most vivid. Much of the book relates Vesey-Fitzgerald's own observations, especially of birds. His account of his initial scepticism about the way snipe make their bleating 'call' during courtship flights (they extend their outer tail feathers, which drum like the sound of arrows in flight as they swoop down), and the way he was eventually convinced, make for one of the most enthralling passages in all ten books.

Over the next half-century the NN covered finches, ants, ferns, orchids, dragonflies; it visited the Peak District, the open sea, caves, the Weald, the Broads. There was a clutch of fascinating titles outside the mainstream: *Pedigree: Essays on the Etymology of Words from Nature* (1973) by the humorist Stephen Potter with Laurens Sargent; *The Art of Botanical Illustration* (1950) by Wilfrid Blunt, avant-garde painter turned Eton art teacher; the polymathic cleric Edward A. Armstrong's *The Folkore of Birds: An Enquiry into the Origins and Distribution of some Magico-Religious Traditions*; and Peter Marren's revealing and waspish account of the stories behind the series, *The New Naturalists* (1995). Its roll call of star authors includes five scientific knights, a dozen Fellows of the Royal Society and one Nobel Prize winner (Niko Tinbergen, *The Herring Gull's World*, New Naturalists Monograph, 1953).

But there are only a handful of female authors in the series, and a list of writers who were conspicuously not invited makes its own comments on prevailing attitudes. Some were just a little too literary to be inside the pale. Kenneth Allsop – more New Wave naturalist than New perhaps – wrote what was in effect a monograph on the little ringed plover, based on his own experiences as a crippled wartime pilot watching the birds in west London's gravel pits (*Adventure Lit their Star*, 1949), except that he composed it as a novel. Jocelyn Brooke's memoir *The Military Orchid* (1948), reflecting on a life in which this rare bloom seemed a constant presence, is both literary and expert, and stands alongside the author's scholarly guide, *The Wild Orchids of Britain* (1950). The poet and critic Geoffrey Grigson's hugely influential

The Englishman's Flora, published by Phoenix House in 1958, could have gone straight into NN alongside *The Art of Botanical Illustration*. John Fowles vowed that he would never write a natural history book. But I think he could have been tempted. When he was not creating seriously European fiction, he was curator of the museum at Lyme Regis, with its incomparable collection of local fossils. He was a remarkable one-off writer on nature and well informed on all aspects. An NN by him on spiders or stones, say, would have been as quirkily knowledgeable as Nabokov on butterflies.

The NN series went into decline during the mid-seventies, and was almost terminated. It's recovering well under a new team of editors, and two recent titles (Rackham, above, and George Peterken, *Wye Valley*, 2008) are outstanding for their literary elegance, and their eclectic mix of ecology, history and polemic. But contemporary titles sell no more than a tenth of the early ones, and the series has not found a style for reaching the huge numbers of people who are members of wildlife trusts and conservation bodies, and whose interest in the natural world is broader and more concerned than the narrow, hobbyist enthusiasms of the immediate post-war world. Maybe we could learn from the Americans. Their natural history books are almost invariably rooted in narrative accounts of work in the field, without any sacrifice of scientific edge. Bernd Heinrich's *Ravens in Winter* (1990) could be a model for NNs of the future. His attempt to unravel the enigmas of ravens' winter behaviour is rigorous to the point of exasperation. But his bewitching descriptions of how he did the work – sleeping with dead goats and cats as raven-bait, trying to stop his crucial chronometers from slowing down in the cold, rigging up an immense sound system to play the ravens' complex repertoire of calls back to them – catch the very stuff of field biology, and the heart of ravenness.

Field biology is not like other sciences. The raw material for its data is chaotic, emotionally charged, wilful, often unrepeatable. Subject and observer alike are battered by weather and accidents and the sheer unpredictability of living things. Its progress can be profoundly influenced by what is happening in the next field, which in turn can be an expression of political and economic decisions made a continent away. The journey from these coal-face encounters to the neatly honed, detached knowledge of the NNs is a long one, involving value judgements and choices beyond the range of the scientific method as it is practised in the laboratory. Maybe it is analogous to the journeys we all have urgently to make in finding a settlement with nature on a troubled planet. The New Naturalist series deserves at least two cheers for providing an incomparable survey, unrivalled anywhere else in the world, of the results of observations of nature in these islands over the past sixty years. But about the making of those observations – about the experience of witness, about the first-hand encounters with nature that are the original raw materials of science and art alike – it says little. We need these insights now like never before.

Some Key Stations in Gilbert White's Life

The following piece was a preliminary treatment, and a kind of storyboard, for a BBC film I made with the historian Michael Wood, entitled Gilbert White: The Nature Man, *2006*

The narrative of White's life, both as a private individual and as the founding father of ecology (and of 'nature writing' – *The Natural History of Selborne*, 1789, is the first example), can best be seen as evolving within a series of formidable constraints and tensions. He was a country boy who went to Oxford, and was seduced by its intellectual and cultural buzz. After his graduation, and then ordination as a priest, he was, first by circumstance and then increasingly by choice, marooned in his remote Hampshire village, longing for the company of his peers. His Christian vocation is barely visible beneath his Enlightenment rationality and early Romantic leanings. And he was afflicted by debilitating coach sickness, which made travel – and escape – taxing. While Joseph Banks was exploring on the other side of the globe, he was out with a lantern, counting earthworms on his back lawn. White's achievements were partly the result of using these constraints as creative opportunities. Emotionally and intellectually, he hunkered down in Selborne, and joined the world outside through writing.

The landscape in which White grew up, and which plausibly shaped his whole outlook, was the muddle of beechwoods, rough commons and sheep-downs which surrounded the village. It's a foregrounded, intricate, inter-dependent system. 'The parish I live in,' he wrote, 'is a very abrupt, uneven country' (a description which would fit his haiku-like journal entries). The beech-clad hill known as the Hanger dominates the village now as then. It reduces the hours of daylight in the street by about three hours, but White used it as a kind of dramatic backdrop against which dramas of migration and breeding were enacted.

A network of hollow lanes wind in and out and round inside the parish, but, typically for a landscape which had pretty well escaped enclosure, formed an essentially *internal* communications system in White's day. They perfectly symbolised his predicament. In bad weather they cut off the village – and Gilbert – from the world beyond. Yet he saw in them something about the resilience of nature and the thrill of ecological change: 'In many places they are reduced sixteen or eighteen feet beneath the level of the fields; and after floods, and in frosts, exhibit very grotesque and wild appearances, from the tangled roots that are twisted among the strata, and from the torrents rushing down their broken sides; and especially when those cascades are frozen into icicles, hanging in all the fanciful shapes of frostwork.' Half a century later, William Cobbett rode down one of Gilbert's hollow-ways: 'Talk of *shows* indeed! Take a piece of this road; just a cut across, and a rod long, and carry it up to London. That would be something like a *show*!' All these landscape features – including the plants that White recorded from them – are still more or less intact.

White attended Oriel College, Oxford, at a time when the cult of natu-ralism in art and literature was growing. He had many poet friends – Joseph Warton (later Professor of Poetry), William Collins, John Scrope – and developed a taste for music and food. It would not be too outlandish to say he mixed in cafe society. The stimulation of Oxford hooked him. (It's commonly assumed that Gilbert's deepest regret was that he could never be full curate of Selborne – the living was with Magdalen College, not Oriel. But it was precisely when he was temporarily occupying that post that his Selborne attachments seemed weakest, and he successfully applied for the post of Junior Proctor at Oriel.)

Aged thirty, fully ordained but without a living, he devoted himself to developing the garden of the family home in Selborne, 'Wakes'. True to his own temperament and developing tastes, he mixed the currently fashionable picturesque style with cutting-edge horticultural experiment. He built a ha-ha between Wakes and the great meadow that stretched up to the Hanger, put up a cardboard model of Hercules and various DIY arbours, forced – and feasted on – prodigious melons. He experimented with novel crops, like wild rice and maize. But most significantly, he began a journal, *The Garden Kalendar*, in 1751, which became the prototype for the *Naturalist's Journal* some years later. The Wakes' garden has been restored recently to pretty much what it was like in the eighteenth century.

His proctorial post lasted one year, and after this White began to cast around aimlessly for a role in life. He took on a few temporary curacies, indulged in a couple of half-hearted flirtations. But he wanted desperately to return to Oxford, and in 1757 applied to be Provost of his old college. He failed, because of a scandal about patrimony and preferment, and this would be the last time he made a determined effort to escape Selborne. He was thirty-seven years old. But something in him had been sharpened. That autumn he wrote an entry in *The Garden Kalendar* quite different from any before, a vignette which combined aesthetic delight with ecological curiosity. '[2 November] Saw a very unusual sight, a large flock of House-Martens playing about between our fields, & the Hanger. I never saw any of the swallow-kind later than the old 10 of October. The Hanger being quite naked of leaves made the sight the more extraordinary.' And four years later comes his first piece of real nature writing, a long eulogy to the crickets whose nest he had found near his house. Nothing remotely approaching the vivid, sensuous attentiveness of this piece exists in eighteenth-century writing on the natural world. He praises their cheerful songs, their industry, their vivid appearances and athleticism. And nothing matches its *respect* for such diminutive creatures: 'a pliant stalk of grass, gently insinuated into the caverns, will probe their windings to the bottom, and quickly bring out the inhabitants; and thus the humane enquirer may gratify his curiosity without injuring the object of it'. It's hard to explain the sudden appearance of this radical essay in the normally restrained pages of *The Garden Kalendar*. Except that Gilbert had seemingly decided that if he could not go to the Mecca of intellectual enquiry, then it must come to him, in the intimate surroundings of what he called his 'green retreat'.

And so begins his compensatory life. He starts to correspond with distant gardeners and naturalists. He answers enquiries from Thomas Pennant for his new edition of *British Zoology*. William Hudson's epic *Flora Anglica* arouses a fascination with botany in him, and he has soon recorded 439 species of plant in the parish (mostly still there, often in the exact places White recorded them). And all this noting and 'epistolary' work is being done with the same lyricism and curiosity and attention to 'minute particulars' that had characterised the cricket essay. He calls it 'watching narrowly'. And one other thing: White begins keeping copies of his letters, despite the labour involved. Already the notion of a parish natural history, done in the precocious, modernist form of a series of letters datelined 'Selborne', is forming at the back of his mind.

There are telling interludes. The 'Festal Summer' of 1763, when three rich and flighty teenage girls from London spend a month partying in the Wakes meadows, and being the subject of curious scientific experiments (including 'electrification') by a bewitched and sublimated Gilbert. And stupendous weather 'meteors' – droughts, terrible black winters, floods – all recorded in his spare, meditative prose: 'On this cold day about noon a bat was flying round Gracious street pond, & dipping down and sipping the water, like swallows, as it flew: all the while the wind was very sharp, & the boys were standing on the ice!'

In 1774 he read before the Royal Society his essays on the swallow, swift and martins, the parish familiars he watched so assiduously from his study in the Wakes. As essays they are masterpieces, not just because of the exactness of the observation and the warmth of the prose, but because, without being in any way anthropomorphic, they have resonance. They are about livelihood and play, about industriousness and family life, about the preoccupations of all social creatures. And they become richer when you understand Gilbert's own social position, as a childless bachelor rooted in his niche. Writing of swifts just prior to their migration: 'we could not but wonder when we reflected that these shiftless beings in a little more than a fortnight would be able to dash through the air almost with the inconceivable swiftness of a meteor; and perhaps, in their emigration must traverse vast continents and oceans as distant as the equator'. Many writers assume, wrongly, that Gilbert did not believe in migration, but thought that summer birds hibernated in ponds. He wondered if a few might (and sometimes searched the winter commons for them), but it was an understandable yearning, a hope that a few of his festive summer housemates might stay close to him during the winter. These four essays must stand as the founding texts in literary ecology, the first writing to suggest that other creatures not only shared common physical space with us, but psychological space too.

In 1783, severe weather hit the village. The volcanic eruption of Skaptarjokull in Iceland ruined the summer: 'The sun, at noon, looked as blank as a clouded moon, and shed a rust-coloured light on the ground, and floors of rooms ... All the time the heat was so intense that butcher's meat could hardly be eaten on the day after it was killed.' (Gilbert often 'measures' meteorological conditions by domestic details like this.) 'Siberian' snowfalls block up the lanes. But slowly Gilbert assembles his antidote to isolation, editing together his correspondence and journal entries and his Royal Society papers into a book which would celebrate the parish's seclusion and community, rather than resent them. During the late 1780s White's extended family became a publishing cottage industry, with nephews transcribing letters, niece Molly in London doing proofreading and Gilbert himself – between spells working on his index – running a kind of crèche in the Wakes. 'In return for your care about my brat,' he wrote to Molly, 'I have pleasure to inform you that your boy is perfectly well and brisk.' The end product, *The Natural History of Selborne* (published in December 1788) was likened by Virginia Woolf to an epic novel, with Selborne 'set solidly in the foreground', a vast cast of characters of all species arriving and interacting, and then the whole stage widened to admit the immense forces that shape climate. The 'narrow' view of the parish becomes a glimpse of a whole world.

Yet curiously, perhaps the most revealing note Gilbert wrote never made it into the finished book. In 1784 the Frenchman Jean Blanchard's balloon passed over Selborne, and Gilbert watched it in something close to rapture, his excitement and envy finally leading to an acceptance of his own stationariness. In his journals this vision of the scientific future is mapped and measured by the ancient and familiar geography of the village: 'From the green bank at the SW end of my house saw a dark blue speck at a most prodigious height, dropping as it were from the sky, and hanging amidst the regions of the upper air, between the weather-cock of the tower and the top of the maypole ... For in a few minutes it was over the maypole; and then over the Fox on my great parlour chimney; and then behind my great walnutt tree ... To my eye this vast balloon appeared no bigger than a large tea-urn ... I was wonderfully struck at first with the phaenomenon; and, like Milton's "belated peasant", felt my heart rebound with fear and joy at the same time. After a while I surveyed the machine with more composure, without that awe and concern for two of my fellow creatures, lost, in appearance, in the boundless depths of the atmosphere! ... At last, seeing with what steady composure they moved, I began to consider them as a group of Storks or Cranes intent on the business of emigration.'

Ronald Blythe

First published as a preface to Ronald Blythe, Aftermath, *2010.*

I first met Ronnie, as he graciously allows us to call him, in 1969, just after the publication of his classic, *Akenfield* – a remarkable book, an intimate portrait of an English village which became an international bestseller. It comprises first-hand recollections by some fifty inhabitants of an agricultural parish in Suffolk, whose memories range from horse-tending in the Great War to modern district nursing. Blythe described it at the time as 'a strange journey in a familiar land'. The familiar land is Deep England, here not at all mythological but part of the lived experience of seventy-five years of tumultuous change in the countryside. The strangeness lies partly in the way the characters' rich stories refute rural stereotypes, and partly in the language of their accounts, which is compellingly authentic yet touched at the same time with an eloquence and gravitas that sound immemorial. As someone who had dreams of becoming a 'country writer' myself, it knocked me for six, especially its lyric voice. I was working in the education department of Ronnie's publishers, Penguin, at the time, and I used this tenuous connection as a way of getting to meet him, hoping privately that I might learn how it was *done*.

I hadn't needed an excuse, as it turned out. When I arrived at his cottage he was as generous as he always is to the legions of admirers, invited or not, who have made the pilgrimage to Suffolk to meet him. He had just come in from clearing out the garden pond, but had found time to make me a cake. The conversation, as I recall, ranged dizzyingly through the vernacular archaeology of ponds, the student revolts of the previous year, the ominous threat of Dutch elm disease, and the edition of Hazlitt he had just been working on. I asked how he had so perfectly caught the voices of *Akenfield*. Had he used a tape recorder? Was the editing difficult? He sounded courteously surprised. There was no technology involved. He simply listened. And then wrote down what he remembered. What he was too modest to mention, of course, was that these rememberings were acts of imaginative reconstruction, underpinned by a lifetime of exploring the local landscape, of hearing and absorbing the subtle rhythms of local talk, and above all of reading – everything from the Song of Solomon to village signposts.

In the four decades that I've been privileged to know him as a friend, Ronnie has not lost a jot of that intense curiosity and unselfconscious alertness to the connectivity between life's elements. It's not just that his 'ordinary' life – talking to cats, preaching sermons (he is a lay canon), chastening the nettles – is seamlessly connected to his writing life, but that they are the same life. Writing (and reading) for Ronnie is simply conversation, a form of extended gossip about the stream of one's consciousness and activities. There's a passage in his *Wormingford Trilogy* (a collection of short but vivid seasonal essays based on his regular column in the *Church Times*) where he describes how a dragonfly settles on a hymnal he's reading in the garden. 'The dragonfly's wings are colourless and translucent, and I can read Binchester and Yattendon through them.' The flesh becomes word.

When Ronnie inherited the painter John Nash's old house, Bottengoms, in 1977 (he had cared for the artist during his final years) he was at one of the focal points of Suffolk's cultural history. The farmhouse in Wormingford is only a few hundred yards from John Constable's River Stour. Over on the coast is Aldeburgh, home of the poet George Crabbe and later of Benjamin Britten. Ronnie had worked for the Aldeburgh Festival in its early years. Every time I went to stay with him we would walk the neighbouring parishes, or I would drive him on mazy tours further afield (he has never owned a car). He knew the houses – and gave the strong impression that he knew them personally – of every local artist and writer, living or dead, and where they had found their furniture. At Benton End there had been the painting school run by Sir Cedric Morris, who also bred irises, some of which were now growing in the garden at Bottengoms. In the village churchyard of Boulge he showed me the grave of the poet Edward Fitzgerald, translator of *The Rubaiyat of Omar Khayyám*. Ronnie had been there when the Shah of Persia came over with a cutting from the rose that grew on Omar Khayyám's tomb, to plant on Fitzgerald's. In Polstead, he told me the story of the infamous Red Barn murder in 1827, and how Maria Marten's killer had been publicly hanged, his body flayed and dissected and his skin used to bind a copy of a book about the murder. That he could seamlessly join this tale with the modern presence in the village of the crime writer Ruth Rendell was part of his genius as a storyteller. Once he guided me to a remote churchyard near the coast where there were the unlabelled graves of the last, hushed-up outbreak of plague in England in the twentieth century. When we got home for tea, he would scurry about for books and prints to show me, do outrageous impersonations of John Nash and E. M. Forster, and explain what it was like to be an aspiring writer in rural East Anglia in the 1950s, when he was working at the library in Chelmsford, and inviting writers up to speak to a literary club he had started.

So though this new collection of his occasional writings, *Aftermath*, consists in the main of book reviews and introductions to other writers' work, it is not in any highbrow sense 'bookish'. The writers featured in it – they range from Hilaire Belloc to Jan Morris – are, so to speak, friends at a distance, and he suggests that their everyday lives and emotions are as connected with their books as his own. It's no surprise that *Aftermath* begins with reflections on journals, the literary form in which the bits and pieces of a writer's life are most intimately entwined. Nor that Ronnie's own journal is no mundane account of his daily round (those details are indelibly recorded in an extraordinary memory bank) but a writer's daybook, in which he sets down his responses to the day's reading and reflection. 'Put everything down,' a poet friend once advised him. 'The total will surprise you.' And surprise us too. Here are his thoughts on the journals not just of literary heavyweights like Dorothy Wordsworth and Samuel Johnson, but also another Johnson, William, the Eton schoolmaster who wrote the 'Boating Song' – and also a cryptic confessional about abandoning teaching and the ethereal attractions of boys. More down to earth are the diaries (or perhaps they should be called workbooks) of William Dowsing, sent by Lord Manchester to 'cleanse' the churches of Suffolk in the 1640s. Dowsing was a local farmer from Laxfield, which only partly explains why his daily record of wrecking crucifixes and carved angels is as chillingly matter-of-fact as an inventory of cattle sent for slaughter.

Ronnie quotes with approval Virginia Woolf, whose view of a journal was that it should be like 'some old desk ... in which one flings a mass of odds & ends without looking them through. I should like to come back, after a year or two, & find that the collection had sorted itself out & refined itself & coalesced, as such deposits so mysteriously do, into a mould, transparent enough to reflect the light of our life ...' In one sense this is how Ronnie's essays and excursions seem to have been organised, except one knows a very writerly consciousness is working away behind them.

It is bewitching to trace how the narratives wind effortlessly this way and that, joining reading and first-hand experience, compressing past and present so that an event or an insight from a thousand years ago is as real as yesterday evening's Bluebell Party in the woods at Tiger Hill, when, incidentally – except that nothing is incidental in Ronnie's stories – a Stone Age axehead punctured the tyre of the Range Rover he was being driven home in. These kind of synchronicities happen to him – or are just acutely perceived – repeatedly. The section on country life ('From the Headlands') contains an essay that is eventually about writers such as Henry Rider Haggard, and the literature of the great agricultural depression in East Anglia. But the route by which Ronnie arrives at this theme is his unique stylistic signature.

It begins with the village schoolchildren, who sometimes have informal classes in his house, bringing armfuls of newspapers down from the loft. He sits down with them to read 'the juicy bits'. Stanley is emerging from the black depths of Africa. Magic lanterns are £7. 17s. Three thousand acres of top-quality farmland near Beccles are to be let for shooting. The letters columns complain about 'irregular marriages'. Ronnie remembers that his own grandmother would have been getting wed just about the time a build-er's apprentice was heaving all this newsprint into the loft to spread between the joists, and was on her way to live in a hamlet called Cuckoo Tye. His grandfather died when Ronnie was six, but he remembers him as a 'quizzical authoritarian with sky blue eyes'. Grandmother was born in 1860 ('years before Hardy had written a word' – a typical Blythean reference point) and lived long enough 'to glimpse our first television set, a sturdy affair with the lines of a fruit machine' (another). Her longevity fascinated him, and led eventually to the writing of a classic book on old age, *The View in Winter* (1979). And so we move to another nonagenarian, Major Bush (alias the writer Michael Home), who lived in the Breckland, once East Anglia's great heath, where there were always gypsies about. Grandmother had once seen one of their vardoes burnt after a death. Moving about so much, had the Romanies noticed the Victorian 'flight from the land'? – and at last we have arrived at the notional subject of the essay. But this overture – not much more than a thousand words long – has taken us across East Anglia, through Ronnie's family, deep into the sources of his inspiration, and graphically sketched in the tone of Victorian country life.

This serpentine narrative isn't random rambling, nor, as chains of associations go, entirely 'free'. It slips effortlessly between the key poles of family and friendship and territory. This is how real lives and thoughts move, not in organised, logically unfolding patterns. Ronnie's gift is to so faithfully reflect this wonderfully connected muddle that what is some of the most elegant prose in modern English reads like conversation.

It is hard to credit that one lifetime can contain so many books read, so many writers understood and remembered (Bunyan, Clare and Wodehouse are all here, and give an idea of *Aftermath*'s diversity), such 'a mass of odds & ends' knitted together. But his work has grown like an ecosystem. A journal entry from the Great War years by Virginia Woolf (who lurks throughout Ronnie's work like a seductive shadow), expressing her incomprehension that the conflict, 'this preposterous masculine fiction', keeps going a day longer, is echoed in his moving and often angry writing about the poems and letters that emerged from the two world wars. The twentieth century's casual sacrifice of its young (and old) is one of his greatest pains. He deplores brutality of all kinds, especially that which lurks under that deferential rural cap, 'the way we do things here': the hanging of poor nineteenth-century fieldworkers for the sake of the pheasant; the militaristic frenzy of women from the Big Houses prior to the Great War; the callous indifference to nature of modern industrial farming. He continuously gives the lie to the idea that a country person must, *ipso facto*, be a conservative. The rural idyll, he says 'slips in blood'. He did his stints at peace vigils during the tense Cold War stand-offs of the 1980s. And though he applauds the way in which East Anglian writers such as Haggard, Crabbe and the novelist Mary Mann exposed the stark facts about the desperate lives of the rural poor, he will not dismiss them as 'units of the brutish, measureless, human undergrowth'. He celebrates instead their knowledge and skill, their survival instincts. Mary Mann, paradoxically, catches their 'actual voice, thus undermining her sense of their hopelessness, for it comes through witty and strong'.

Which brings us to William Hazlitt. The essay on the eighteenth-century radical essayist shows Hazlitt not so much as Ronnie's hero, as his alter ego. The two have much in common: they are self-educated countrymen – 'village intellectuals', passionate liberals, Romantics in love with language and with their fellow humans, of whatever station. But Hazlitt was burdened with baggage Ronnie has mercifully escaped. He was bitter, subject to mood swings and outbursts of violence, intemperate in just about every possible way and incapable of not wearing his heart on his sleeve.

By contrast, Ronnie's work, though deeply personal and often autobiographical, is intensely private. Do not expect disclosures or revelations. The committed and practising Christian never evangelises outside the pulpit. The man who has written so penetratingly about the impact of war on the ordinary soldier says nothing about his own military service in World War II. The sensitive interpreter of other writers' love lives is silent on his own. If you were to ask him why, the answer might well be that robust East Anglian rebuff to nosiness, 'Never you mind!' Time enough for that when the biographies are written. But I sense civic courtesy as well behind this diffidence. Over the past half-century we have been slavered with self-indulgent memoirs and egotistical confessionals, the literature of the 'me' generation. Ronnie's personal writing offers something far more valuable and noble: the literature of 'us', where the 'I', so to speak, becomes the eye, fascinated with the world beyond itself.

Coleridge once recommended that introspective writers would be better off going for a tramp through the woods than contemplating their tortured egos. Ronnie's walking is a window on his life and an analogue of his prose. Unlike many of his subjects he has not exactly 'stayed put' but taken the

East Anglian tradition of 'getting about' across the Atlantic, and been talked to by the Sheriff for walking a little too freely on the freeways. But it is perhaps his account of night-walking at home (in *Word from Wormingford*) with its sense of history and continuity and transcendental detail that might serve as an epigraph. 'The church tower is a charcoal stump, just as it was during the summer nights that followed the Conquest ... Gravestones are legible and there are dense scents. Young rabbits are dining off a wreath and other unidentifiable creatures rustle and fidget. Everywhere, it is all so perfectly interesting that one might never go to bed.' The beat of the heart is here as well as the feet. Out in the East we regard Ronnie as our tribal storyteller, plugged into the common stream of inquisitive conversation that joins us as a species.

Cellmates

First published in Slightly Foxed, *2012*

As I remember it, *Vole* was already up and running when Lewis Thomas appeared in our midst like some ecological genie, a combination of sage, comedian and word conjuror. It was 1977, and Richard Boston, founder of the magazine, arrived at an early editorial gathering bearing a copy of Thomas's *The Lives of a Cell*, with the clear message that it was required reading. It had recently been awarded, unprecedentedly, *two* US National Book Awards, one in the Arts category, the other in Science, and been described in the *New Yorker* as a 'shimmering vision'. Thomas was something of a vision himself, as improbable as a tortoise with wings. He was a scientist who was not only literate but poetic. He was a polymath, as learned about etymology as entomology. He had written a meditation on listening to Mahler's Ninth Symphony, whose subtext was a troubled attack on his country's nuclear weaponry. And somewhere he found time from his day job as a medical academic to contribute a left-field column called 'Notes of a Biology Watcher' to the *New England Journal of Medicine*, on subjects that seemed as bizarrely disconnected as moth pheromones and Bach's 'St Matthew Passion'.

But of course they aren't disconnected, except in the most literal sense, and it was Thomas's genius to conjure some thirty of these essays into a coherent treatise full of reverberating knowledge and sublime prose. Against the grain of the times, he shunned environmental doom and New Age waffle, and celebrated instead the tenacity of life and the way, specifically and allegorically, it all joined up. And greatly to our magazine's point, it was wickedly, surreally funny. We fell on it as if we'd been handed down a house-style bible from on high.

Vole had been established to explore, in a necessarily more quotidian fashion, similar themes as Thomas. Seventies Britain, despite the rising interest in environmentalism, had few regular publications which explored it in an accessible but serious way. We hoped our magazine might fill the gap. It would be an eco-literary monthly, radical, satirical, and iconoclastic of environmental piety on the one hand and forelock-touching, 'country writing' on the other. The title, shortened from 'The Questing Vole', had been inspired by the famously gloopy phrase Evelyn Waugh attributes to his nature columnist hero in *Scoop:* 'Feather-footed through the plashy fen passes the questing vole.' The cartoonist Posy Simmonds had designed the logo, a vole with a quill pen behind its ear. The result was a hard-to-classify hybrid between *Private Eye*, the *Whole Earth Catalog* and the lighter sections of the *New Statesman*. Under its banner Boston assembled a starry and eclectic class of contributors: Terry Jones (who provided the start-up money), Miles Kington, Gillian Darley, Paul Foot, Jeremy Bugler, Simmonds, Ralph Steadman, Richard Ingrams and Richard D. North (there were so many contributors with the forename Richard that there was once a special issue where all the writers listed on the cover were so named), who before his mysterious metamorphosis into the intellectual voice of environmental contrarianism was the cheerfully anarchic contributor of a bicycling column entitled 'Volocipede'. *Vole* was full of excruciating wordplay like this, which I like to think Lewis Thomas, in his etymologist's robes, would have applauded as examples of the frequently mutant evolution of language.

His own language was anything but mutant. It was exact, idiosyncratic and often heart-stoppingly beautiful – never more so than in the book's key passage, where he reimagines that step-changing first photograph of earth from space: 'Viewed from the distance of the moon, the astonishing thing about the earth, catching the breath, is that it is alive. The photographs show the dry, pounded surface of the moon in the foreground, dead as an old bone. Aloft, floating free beneath the moist, gleaming membrane of bright blue sky, is the rising earth, the only exuberant thing in this part of the cosmos ... it has the organised, self-contained look of a live creature, full of information, marvellously skilled in handling the sun.'

'Has the look of' is Thomas's signature phrase. Bees building a hive 'have the look of embryonic cells organising a developing tissue'. Language itself has the look of an evolving organism. This is his favourite trope for introducing startling insights (or maybe foresights) into the patterns of life. He rides the high and perilous ridges of grand metaphor faultlessly, never tumbling into the crevasses of sloppy anthropomorphism or absurd fantasy.

In an essay entitled 'The Music of *This* Sphere' he imagines the combined sounds of all the world's organisms as symphonic, as a 'grand canonical ensemble'. If we were able to hear them all together – the proboscal flutings of death's-head hawkmoths, the rhythmic drumming of schools of molluscs, the distant harmonics of midges hanging over meadows in the sun – 'the combined sound might lift us off our feet'. And then he drops in the possibility that whales may hear their own convoluted melodies again after the songs have circumnavigated the globe, and work critically on new variations ... Thomas is not saying that any of these events has actually occurred. They are possibilities, pictures of feasible worlds – possibility of course being the template through which evolution proceeds.

He explores an analogous connectivity in 'Vibes', about communication by smell (or at least by vaporous chemical messaging) and has a delicious vision of the earth – again seen as a kind of giant cell – kept up to speed by a worldwide web of scent: 'In this immense organism, chemical signals might serve the function of global hormones, keeping balance and symmetry in the operation of various interrelated working parts, informing tissues in the vegetation of the Alps about the state of eels in the Sargasso Sea, by long, interminable relays of interconnected messages between all kinds of other creatures.' This was another what-if metaphor, not a scientific theory – but only to the extent that Thomas was extrapolating known phenomena (e.g. the complex linkages between trees in forests) onto a much larger stage. Metaphors often do have literal counterparts. The chimeras of mythology – centaurs, griffons, phoenixes – have real world analogues, not least in our own cells, which are hybrid communities of small animals and plants living independently in these minute bags of fluid.

We didn't explore many of Thomas's themes directly in *Vole*, but his constant emphasis on symbiosis, repeating patterns, weird synergies and improbable connectivities seeped into our poorer – but porous – imaginations. We devoted most of one issue to an elaborate spoof that satirised two targets – New Mysticism and new megacompanies – with one shot. British Leyland, we exclusively revealed, was a latter-day Druidic cult, and the motorway system a network of modernist ley lines. The map which showed them linking the sacred sites of big oil culture, such as Luton Airport and the BL plant at Coventry, via the holy wells of service stations, was alarmingly convincing. The vintage 'Slow' issue, which had features on snails, steam trains, the 'slow egg' (cooked by being wrapped in black paper and left out in the sun), the virtues of procrastination and, of course, on sloes, predated the Slow Food movement by nearly a decade.

Looking back after forty years, some of this student japery makes me cringe. But a lot of it echoed Thomas's impish habit of shaking the living world like a kaleidoscope and seeing what interesting constellations revealed themselves. The most glorious chapters in his book are on human language, which 'has the look of' an evolving ecosystem. His application of a biologist's thought processes to this system, which may be a defining characteristic of our species, is revelatory. 'We have DNA for grammar, neurons for syntax. We can never let up; we scramble our way through one civilisation after another, metamorphosing, sprouting tools and cities everywhere, and all the time new words keep tumbling out.' And when they 'unfold out of old ones, the original meaning usually hangs around like an unrecognisable scent, a sort of secret'. He traces the family tree which emerged from the ancient root *gene*, meaning beginning, giving birth. It morphed, for example, into *gecynd* (Old English) meaning kin or kind. 'Kind was at first a family connection, later an elevated social rank, and finally came to rest meaning kindly or gentle. Meanwhile, a branch of *gene* became the Latin *gens*, then gentle itself; it also emerged as genus, genius, genital, generation and generous; then, still holding on to its inner significance, it became "nature" (out of *gnasci*).' What a beautiful arabesque, and how wonderfully it catches Thomas's unfashionable and inclusive optimism.

Vole thrived mightily for a couple of years, then declined and finally became extinct in 1982. If Thomas had been describing its collapse he might say that it had the look of a colony of social bees where the workers had lost track of how to find nectar. Basically, we hadn't a clue about balancing the books. But most of the colony members evolved to do other related types of honey-making (though Richard Boston died, too early, in 2006). Even Richard North's transmutation would I think have been seen by Thomas (who died in 2003) as benign, by providing developmentally useful provocations for those who disagreed with him.

For myself, I still have my original copy of *The Lives of a Cell*. It is the most frequently reread, quoted, probably plagiarised book in my possession, and remains an inspiring reminder that the course of life on earth is not a predestined tragedy, but a continually unfolding, improvised, extravagantly inventive comedy.

Leaving Tracks

First published in Landscape and Literature, *edited by*
Kay Dunbar, 2005

In our new exploratory relationship with landscape and nature, walking has become the prescribed rite of communion. Walking will set you right, mend your heart – and your head. Walking will help you find your place. Walking makes bonds with the earth. It is as simple, they say, as putting one foot in front of the other ...

For Dr Johnson and George Borrow, both great wanderers in their time, making contact with the world through walking took on a quite literal meaning. Both were dogged by recurrent bouts of depression, and to keep some kind of link with reality they used to touch objects – trees especially – on their way. For William Hazlitt, solitary walking was an aid to contemplation, especially among familiar surroundings. 'I can saunter for hours,' he wrote, 'bending my eye forward, stopping and turning to look back, thinking to strike off into some less trodden path, yet hesitating to quit the one I am on, afraid to snap the brittle threads of memory.' Henry Thoreau also relished 'sauntering', and pondered the possible origins of the word in the phrase *sans terre*, 'without land or home, which therefore, in the good sense, will mean, having no particular home but equally at home everywhere'. He found that he nearly always sauntered towards the south-west, 'where the earth seems more unexhausted and richer'.

When I first began walking seriously round my home in the Chilterns, I was a ritualistic saunterer, too. My family had an assortment of euphemisms for strolling. 'Just off to stretch the legs' or 'Just taking an airing' – as if one were a piece of stale washing. Mine was 'Just going up the Top'. The Top was my songline, a circular beat (to the south-west) that could be done in an hour, but which meandered through a patchwork of woods and green lanes, every scrap of which had profound and totemic meanings for me. When I was a teenager I took this walk three or four times a week, and soon I was following not just my regular route but, like Hazlitt, my own previous footsteps. I would keep to the *right*-hand side of the lane, close to a black-thorn hedge, cut a corner by a big oak tree, and head for home along the edge of the cornfield rather than on the footpath a few feet below it. Maybe this obsessive tendency was a relic of childhood superstition, like avoiding paving cracks. But I also found something reassuring in keeping to my own tracks, a sense of holding the precarious world of adolescence together.

I took possession of that walk, turned it into an emotional refuge. I used it for earnest perambulating debates with friends – and also, absurdly, as somewhere to hide from them. When I was barred from the company of my first girlfriend, I used to bury messages from her under the oak tree. I revised for exams along it, with a pocketful of record cards onto which I'd condensed all the information I was nervous of forgetting. I was, I think, beating the bounds on these ritual plods, checking out both my internal and external *terroir*.

Later, I found favourite walks further afield. Yet staking-out, first-footing, a kind of colonisation in the head, had become a habit. One walk (still further to the south-west) I took whenever the ritual plod round the Top seemed too clogged with stale memories or northern reserve. It passed through a line of riverside water meadows, where I once saw six cuckoos feeding together. Then it crossed a stream, climbed a little and returned along the foot of a beech hanger. The beeches were regularly blown down in storms, and with each new gap there were startling new views of the woodlands beyond. They were cherry woods for the most part, cloaked with white blossom in April, and later ringed by it, when the petals fell and settled round their rims. There was a flamboyant, southerly feel about the people, too. One regular walker did the circuit in climbing boots, carrying a small dog in his arms. Two others I never glimpsed, but I'd found their polished champagne glasses hidden high in the cruck of a streamside oak.

One May here, I saw the actual moment that spring began, and felt for a moment like a witch doctor whose spell had worked. There had been nearly two days of cool heavy rain and I had driven south to try and get beyond it. I sat on a log by the side of the stream and watched the cloud begin to lift. Small bands of swifts and martins appeared, drifting in from the south. Then – it seemed to happen in the space of a few seconds – the wind veered round to the south-east. It was like an oxygen mask being clamped to my face, so sudden that I looked at my watch for the time. It went down in my diary: '6 May. Spring quickening, 4 p.m. exactly.' So the walk developed its own circular logic. Same views, same log, same date (6 May), same clockwise direction. But it was my structure, not the landscape's.

When I moved to Norfolk in 2002, there wasn't a landscape to appropriate in the same way, even if I'd wanted to. The flatness, the treelessness – they give you no cues, no signposts. The arable land has already been spoken for. In the wooded Chilterns – full of cryptic coombes and ancient beeches and capricious scrub – it was easy to establish your own niches. But up here the fieldscape has been flattened out, simplified. The wetlands too are hard to pin down in imaginative charts, not because they have been drained of meaning, but because they are elusive. They shape-shift. They're defined not by landmarks but by flow and nuance. When I walk around the fen down the lane, it is not the same two days running. The paths vanish underwater. The trees collapse, the reed beds are cut. The half-wild Konik horses are browsing in the alder thickets one day, and wading deep among the darting teal the next.

How does the absence of a secure berth, where every track is the road less travelled, affect the influence of landscape on writing? What does it do to the historic role of the walk as *text*, as a medium through which landscape becomes literature? It's no longer as easy, perhaps, as putting one word in front of another.

Many writers used walks to provide narrative structure. John Bunyan's gossipy account of his Pilgrim's Progress mirrored his own wanderings about the Bedfordshire fields and Chiltern heights. John Clare's poems, ramping with immediacy and what Seamus Heaney called 'the one-thing-after-another-ness of the world', conjure up his barely controllable excitement as a boy walker, darting from one side of the lane to the other, 'jumping time away'. One critic reckoned W. H. Hudson's rambling prose 'perfectly echoed the long, slow, unhurried tramping of his feet as he roamed through the gentle southern counties each summer' – though, for me, they echo more the alien footfalls of the census taker.

The patron of literary strollers, though, must be William Cowper. After his first mental breakdown in the 1760s, Cowper went to live with friends in Olney. Like many before and after, he tried to keep his melancholy under control by busying himself with domestic routines, especially gardening and walking. 'The very stones in the garden-walls are my intimate acquaintances,' he wrote in a letter in 1783. His great poem *The Task* broke graphically with the contemporary taste for structured prospect descriptions. It isn't ordered to any particular design, but offers scenes, observations, reflections exactly as they might be encountered on a walk. Scale and perspective are repeatedly shifting, so that one moment a wild flower is in focus, the next a whole cycle of work in a distant field. In one remarkable passage, he plays games with his own shadow on a wall, 'spindling into longitude immense, / In spite of gravity, and sage remark / That I myself am but a fleeting shade, / Provokes me to a smile'.

Was Cowper here close to a different view of walking? A glimpse of it not as a ritual of perambulation, a way of taking possession of the landscape, but as an opportunity for the landscape to take possession of *you*? I'm much taken these days with feral walking. I make no plans, except to begin and end in the same place and follow as few made paths as possible. Instead I tag along on deer trails, head for trees on the horizon, give in to the slightest instinct to change tracks, and hope I'm being guided by cues below the level of my conscious attention. Of course free-form walking is no such thing, and you're being prompted by the scribblings and murmurings of countless other organisms. But it feels less like colonisation than those obsessive, self-gratifying circuits of my youth.

This winter, as often, I'm finishing a book, and not walking much. I gaze out of the study, grasping at any diversion. Deer sidle through the orchard. A female sparrowhawk has just caused mayhem at the bird feeder. I wonder whether these outside events influence the rhythms of what I'm writing. Should I try to shut them out? Or are they the landscape strolling in on me, beating its own bounds?

Outside the other window our builders, Roy and Lee, are working on the walled garden. They're laying bricks in Flemish bond, one pair along, one single across. It's as hypnotic as watching the tide come in. Then, right in front of me, a rat is swimming across the pond. It's seen the prostrate pheasant that fell to a syndicate shoot yesterday, and tries, with a few tugs of the tail, to drag it ashore. But it's too heavy, and the rat swims back to the bank, nibbles some waterweed, as sweet as pie, and bounds like a gerbil across to the bird feeder. It's learning to climb up, but doesn't have things all its own way. Already there's a pecking order: jays and mallards see the rat off; blackbirds just jump in the air.

I feel that somewhere between the rat's opportunist grafting and the measured craft of bricklaying is where I should be steering the words. But I need an airing, and take off for a stroll. There is a Latin tag, *solvitur ambulando*, which means, roughly, that you can sort things out by walking. You often can, except that it may not be the problem you think you wanted sorting. I've ended up on a nearby common. The sun is warm, the grass is riotous with moles, and I just drift about, from hillock to dell. The moles are editing the turf, which is being stetted back by creeping clover roots and the first shoots of plantain and sorrel. I have the distinct sense that the grassland has, so to speak, a *plot*, with its own fuzzy logic.

Back home, it's dusk. Fox-time. In his poem 'The Thought-Fox', Ted Hughes imagines the animal walking the land just as he stalks the page, and entering 'the dark hole of the head'. Outside, the pond level is still rising from some mysterious spring. Immense flocks of rooks fly south-west, as they do every twilight. Lifelines seem to be flowing over and under me. Am I part of them or just watching them? And who is writing this anyway?

The Library as an Ecosystem

First published in Slightly Foxed, *2015*

I like to think we run an open-door policy in our library in Norfolk. That is to say, on warm days in summer the door to the garden is actually open. Anyone's welcome to come in for a browse. Last summer a stoat wandered in, peered dismissively at the modest shelf of my own titles, sniffed about under my desk and then ambled out. Most Julys the house ants – here long before us and so given due respect – pour out from alarming new holes in the floor, march along the tops of my books of botanical art, and shuffle in a lost and desultory way about the carpet, seeming uninterested in getting outdoors for their nuptial flights. But while I fret about the continuance of their ancient lineage, the culling is already under way. Next through the open door come the bolder blackbirds and robins, hoovering them up in front of the shelves.

I think it's entirely proper that the place is a working natural habitat, the word library emerging from the Latin *liber*, which described a bit of bark inscribed with letters, and having secondary meanings to do with the liberties of the marketplace. A collection of books begins physically, I guess, as a kind of landscape. You plan its geological layers, cliffs, niches, antechambers.

Hillocks and book-quakes erupt where the mass gets critical. But the internal organisation is more likely to resemble a magazine advertising supplement than the Dewey decimal system. Space has to be filled, marriages of convenience arranged, beautiful jacket displays sometimes given preference over spinal detail. In our sixteenth-century farmhouse fiction and poetry reside chiefly in the living room, some of them in a disused open hearth. Children's books are halfway up the stairs. An entirely irrational combination of dictionaries and field guides and American non-fiction swarms in a room in the one-time outhouse that I use as an office. Almost everything else lives in what we rather grandly call 'the library'. In here I try to keep a semblance of order by arranging technical and academic books under subjects, and others alphabetically under author, on the dubious assumption that this ordering will mesh better with my memory. One writer I know arranges her travel books in order of the mean temperature of the region they're set in.

This is the point where the library ceases to have the immemorial structure of a landscape and edges towards the rampant wildness of an ecosystem, with an agenda of its own. An old bibliophile's saw is that if you gaze at the spines of your books for long enough, you absorb the contents by a kind of dowsing. I'm increasingly finding I have to gaze at the form of the spines just to locate them. Pattern recognition, an entomologist would call it. I can never remember either the title or the authors of *The Garden, the Ark, the Tower, the Temple: Biblical Metaphors of Knowledge in Early Modern Europe*, but can always find it because it is edged with a distinctive halcyon blue mosaic of Edenic animals and plants, and lives high in the canopy (how apt that 'leaves' are common to both books and boskiness). And for all one's good intentions, orderliness seems to be wilfully disrupted by unintended and provocative conjunctions. It needed no mischievous disruption of the order to make philosopher and fox hunter Roger Scruton's unimpeachably conservative memoir *News from Somewhere* (in which he marries a girl 'whom

I had seen poised aloft in Beaufort colours – like a painted angel in a frame')
stand dialectically next to Alfred Schmidt's severe *The Concept of Nature in
Marx*. But I have no recollection of slightly tweaking the authorial alphabet
so that David Hendy's *Noise: A Human History of Sound and Listening* sits
serendipitously adjacent to Adam Gopnik's lyrical essays on *Winter*, which
included his wonderful evocation of the hiss and spark of Wordsworth's Lake
District skating. The library is having ideas here, before I do.

But it's the seasons of migration that really throw the wild card into the
packed shelves. When I'm on a big project, I usually ferry a week's worth
of likely reference sources from their various winter quarters over to where
I do the actual writing. So armfuls of paper, from thin downloaded journal
articles to eighteenth-century folios cross parts of the garden like herds of
wildebeest, and are set down in the library. There's no room for them of
course, so they go on the floor in strange rows and clusters that I have never
glimpsed just so before. At the moment I am, for a book on plants and the
imagination, trying to make sense of the relation between biological mimicry
and literary metaphor (both cases of 'this stands for that'). The line of
immediately pertinent texts is bookended by a Richard Cartwright painting,
heavily framed in black, of a white whale swimming in a black sea under a
white cloud. The title at one end is Colonel Godfery's classic 1933 monograph, *Native British Orchidaceae* – the only book I have seen where the
dedication page is dominated by a photograph of the dedicatee, Godfery's
beloved late wife Hilda, posed among the foliage in her garden. At the other
end is William Anderson's *The Green Man* (rather archly subtitled 'An
Archetype of our Oneness with the Earth') and I ponder, prompted by their
proximity, whether Hilda was a kind of Green Woman, and the degree to
which foliated hybrids permeate our imaginations.

This group of books, in ecological jargon, would be called a 'guild', a
set of organisms that all fill similar roles or do analogous jobs. Tree species
that grow in swamps form a guild. So do insects with a taste for carrion.

My current books are a guild in that they are all loyally devoted to a single purpose: helping me get this disorderly chapter – currently throwing off sideshoots like a bindweed – sorted. But they are also, ecologically speaking, an 'association', a community of species which live together in one place, usually in some kind of mutually beneficial relationship. The insects, plants, fungi and bacteria that make up a termite mound – an apt analogy to my literary piles – make up an association.

A few groups of books, often for entirely sentimental reasons, have been taken out of the library's chaotic but well-meant order and encouraged to associate. I have one small bookcase – more anthill than termite condominium – entirely devoted to the works of John Clare and Ronald Blythe. I thought they would be happy nestling up together not only because Ronnie has been such a champion of and prolific writer about Clare, but because they both grant huge significance to the places where things dwell – and what principle is more important in a library? One of the choicer items in the Blythe section is an inscribed edition of the lecture he delivered at the Royal Society of Arts' Nobel Symposium in 1980, entitled 'An inherited Perspective'. In it he describes his shock at glimpsing John Clare's village of Helpston from an express train, 'first the platform name, and then the niggard features of one of the most essential native landscapes in English literature'. He had not even realised it was on a railway line.

Margaret Grainger's collection of *The Natural History Prose Writing of John Clare* (1983) quotes Clare's graphic and curmudgeonly account of the sixteen species of wild orchid that grew round Helpston. Denouncing botanical science as the 'Dark System', swatting aside even the Bard himself, Clare proclaims the natural *rightness* of vernacular names: 'let the commentators of Shakspear say what they will nay shakspear himself has no authority for me in this particular the vulgar wereever I have been know them by [these names] only & the vulgar are always the best glossary to such things'. What emerges is not just a lexicon of popular names but a vivid list of the orchid's *addresses.* The early marsh orchid, one of Clare's 'cuckoo buds', was 'very plentiful before the Enclosure on a Spot called Parker's Moor near Peasfield-hedge & on Deadmoor near Sneef green & Rotten moor by Moorclose but these places are now all under the plow'.

Clare is by no means the only writer to have 'named' and ascribed the identity of a plant by its dwelling place. When Wordsworth's famous 'Primrose of the Rock', once an anchor in his Lakeland life, degenerated into a banal religious symbol, his sister Dorothy nailed it firmly back to earth in her subtitle for William's final verses on the flower: 'Written in March 1829 on seeing a Primrose-tuft of flowers flourishing in the chink of a rock in which that Primrose tuft had been seen by us to flourish for twenty nine seasons.' The great Romantic scholar Professor Lucy Newlyn thinks that 'such elaborate specificity suggest she saw the poem as belonging to the "Inscription" genre, normally used to commemorate connections between people and places'. I first had contact with Lucy when she sent me a copy of her delightful paper on literary glow-worms, which should by rights be living close to the book I have just quoted from, *William and Dorothy Wordsworth: All in Each Other* (2013), but which for reasons of its own, has migrated to the outhouse, and a sub-sub-section of 'special gifts' . . .

You may begin to see where I am going in this round-the-rocks digression. Inscriptions, connections, authors, places – all prime librarious issues. Back in the book room the real problem comes when those briefly privileged titles have to go back 'where they belong'. Some hope. They've acquired new acquaintances, new guild partners, so their belonging has become a serious existential problem. Frankly, I'm often tempted to give up any pretence of order, and arrange the whole lot by free association, so that the library would become a leaf-space as vivaciously mobile as a rainforest canopy. I'd never find anything I wanted, but just imagine the new symbioses and connec-tivities I'd discover by – to only marginally misuse the term for what drives evolution – natural selection.

PART TWO

Considering The Lilies

As a young scribbler, I never gave plants a thought. Like most proto-naturalists, I began as a boy birder, poised between obsessive list-maker and airy romantic. I logged the dates of the first chiffchaff, and watched the swifts arrive back at the parish church in May. My first intellectual interest in, and writing about, the natural world was the history and workings of the land-scape. I found plants decorative but inanimate, their lives embedded in a language (stigma, petioles, ligules) I found hard to penetrate. But over the next forty years more than half my books had plants in a starring role, suggesting that I quickly recovered from that youthful blind spot.

With hindsight, an isolated incident from my teenage years pointed up what would eventually turn my head. One summer afternoon, my sister Pat, my mentor in all things nature-related, came home from a walk by the canal with news of a plant which didn't seem to be in her copy of Rev Johns's *Flowers of the Field*. Always principled, she hadn't grabbed a sprig, but described it as having curious nasturtium-like flowers in freckled orange. Eager to score one over my elder sibling, I went off to the spot she'd outlined, found the cryptic tuft without too much trouble, memorised its details and went to the public library to look it up in a more comprehensive guide. It was orange balsam, scientific name *Impatiens capensis*, an immigrant from the USA. Having something of a penchant for Latin, I was intrigued to read that its family name *Impatiens* derived from the irritability of its ripe seed pods, which burst open when disturbed by touch or wind, propelling the seeds metres into the air. Decades later I learned that the species once had the delicious vernacular name of 'quick in hand', and was first noted in the wild in Britain by the philosopher John Stuart Mill, who found it while botanising by a stream in Surrey in 1822. I'd read Mill's utilitarian philosophy at Oxford, and was tickled by the thought that he should have discovered a plant so totally useless to humankind, but so canny in ensuring its own legacy.

The idea that plants might receive some animation from their entanglement with human culture resurfaced when I was exploring the marshes on the north Norfolk coast in my early twenties. I've outlined in this section how the discovery that there was still an active tradition there of eating shoreline plants led to my first nature book (*Food for Free*, 1972). I'd chanced on Geoffrey Grigson's *An Englishman's Flora* at much the same time, and was enraptured by its lush prose and stories of how wild plants had entered our language, larders and sense of landscape. Learning the local names of marsh samphire (poor man's asparagus, sparrowgrass, sandfire) and the local lore concerning the harvest (best after the shortest day, must be washed by every tide) and how it was cooked and eaten (on the stalk, drawn between the teeth) was a whole new way of engaging with the character of a vegetal companion. But sometimes, watching the tide go out over acres of mudflat bristling with its bright green stems, something more fundamental and foreign captured my imagination. I caught the sense of a life quite separate from us, an organism with its own agenda of survival which had no need of cultural usefulness to create an identity. These two ways of regarding plants, as deeply embedded in our everyday lives and as autonomous beings telling their own stories, have debated with each other in all the plant-orientated books I wrote subsequent to *Food for Free*, especially in the title that followed – *The Unofficial Countryside*, about the feral life of urban areas.

In the year that *Food for Free* was published I was introduced to the photographer Tony Evans by the editor of Nova, with a view to producing a words-and-pictures feature on the fortunes of Britain's wild flowers. The long-term results were a close friendship that lasted until Tony's untimely death in 1992, and *The Flowering of Britain*, a social and cultural history of Britain's wild plants (published in 1980, and made into a BBC TV documentary the same year). The project involved us spending six unforgettable summers on the road together, and I've told the story of those expeditions in a chapter in *The Cabaret of Plants* (see below). I explained how Tony's way of looking changed my own perspective on plants, how he developed still life into ecological narrative. One story in particular seems to sum up his vision of the world. We are in Bradfield Woods in Suffolk, contemplating oxlips ... 'And as we talk, Tony sizes up the flowers around us. He chooses a luxuriant clump, with the flowers in a pyramidal, almost tent-like canopy above the leaves. He anchors the tripod, like a stout umbellifer, and takes the camera down to woodland floor level. After a while he finds in the viewfinder the narrative picture he will take – a dozen or so flower spikes, deeply grounded but rising like a sheaf in the left foreground, tangled up with dead twigs, young meadowsweet flowers and a protective tussock of grass. In the background, caught by the wide-angle lens, is a fragment of cut-over coppice, and beyond that new hazel stems, still leafless, shearing towards the clear spring sky.'

Over the years other nuances of plant meaning emerged. I was invited to write a commentary for a cache of Victorian flower paintings discovered in a manor house loft in Gloucestershire. *The Frampton Flora* (1986) was the work of a bevy of sisters, aunts and cousins, remarkable for their sensitivity to the way in which plants helped create a sense of place. And in 2005 to write an account of the Eden Project, that theatre of plants in Cornwall, which, always too clever for my own good when it came to titles, I cryptically dubbed *Fencing Paradise* (2005). I wrote a biography of the beech (*Beechcombings*, 2007) which turned into an elegy for the wooded hills of the Chilterns where I'd grown up and a meditation on how our models and stereotypes of trees collided with their 'natural' identities and life stories. The culmination of the cultural strand was *Flora Britannica* (1996), a kind of Domesday Book of Britain's wild plants, an attempt to log what I called their 'modern folklore'. Among a host of unclassifiable recollections sent in by members of the public were records of contemporary Christmas customs, landmark trees, children's plant games, intensely personal plant associations. Collectively they built up two interpenetrating geographies: the British people's immensely diverse affections for (and prejudices about) plants, their imaginative interpretations, their sensitivity to locality and the calendar; and the plant world's echoing of all these qualities in its inventive oddity and local variety. I was gratified that one of the contributions recalled my primal botanical experience of the explosive power of the *Impatiens* family. A group of youngsters in Cumbria had learned the trick of reading Indian balsam's pods for the slight change in colour and texture that indicated they were ripe for bursting, and had held a seed projection Olympics. The winner notched up twelve metres.

In 2007, Andrew Franklin at Profile Books asked me whether I thought a book on the cultural history of weeds might be viable, and who might be a suitable author. I replied, feeling more sure than I ever have about a book project, that, yes, it was viable and that I must be the author. *Weeds* (2010, and translated into ten languages) marked something of a turning point, and a paradox. It was about a category of plants that, both conceptually and ecologically, was a human creation. But the story turned out to be more about plants as creative, autonomous beings. I began to think about the extent to which our appreciation of them was framed in human terms. Ideas about their beauty, for example, seemed to be principally about us. Was their visual appeal – the curve of leaves, the delicate shades of blossom – meaningless except as a sop to our pleasure centres? Might there be less human-centred registers of beauty, which expressed the grace with which plants lived in the world, their relations with other organisms, a range of senses that far exceeds our own? How might we approach plants not just as objects in our lives but as subjects in their own, without reducing them to the green hominids they are in fairy stories or Victorian moral parables? How might metaphor and allusion reflect the otherness and autonomy of vegetal existence, while celebrating the fact that plants and ourselves are part of the same living universe?

These were the questions I considered in the book that followed *Weeds*, *The Cabaret of Plants* (2015). I subtitled the book 'Botany and the Imagination', and explored how the strangeness of the vegetal world has always supplied its own challenging meanings for life and death, sociability and language, and, most radically, intelligence (PI), and how these have stimulated human thinkers from Ice Age rock artists to twenty-first-century behavioural scientists.

Round The World in Eighty Dishes

First published in Slightly Foxed, *2016*

I slipped into the world of Lesley Blanch's swashbuckling cookbook, *Round the World in Eighty Dishes*, before I'd even heard of it. It was the early sixties, and I was on my first visit to Paris with a gang of friends from university. The city was sizzling in a July heatwave, and our host took us to an Arab quarter near Saint-Michel, where we saw something extraordinary to our English eyes: people not just eating in the street but *cooking* in it. The North Africans performing legerdemain on their charcoal braziers seemed more like a troupe of fire-eaters and jugglers than cooks. One was frying green peppers with what looked like small sausages in a large pan, and adding and pounding chopped tomatoes. With a flourish, he tossed half a dozen eggs onto the bubbling surface. The result was green and gold and scarlet and swirly – a Fauvist painting in a pot, and I thought it the cleverest and weirdest food I had ever seen. I didn't know its name, and never got to sample it, but I resolved to try and recreate it when we arrived at our holiday destination, a villa on the Spanish coast near Barcelona.

It was my first trip to Spain too, and the food culture was a revelation. I was an adventurous eater back home, working my way through the Asian cuisines already available in English restaurants. But I was a naive cook, with no real sense of how cuisines 'worked' and how their dishes were an expression of a whole culture. The Catalan coast in high summer changed all that. We were taught how to eat soft shell crabs whole, to be unafraid of calamares. The beaches were full of pop-up cafes where, in the same spirit as the Parisian alfresco food stalls, immense paellas were cooked continuously over braziers. Those days are forever fixed in my mind in the mixed aromas of saffron and Ambre Solaire. One day, driving in the hills, we were stuck for hours behind a broken-down lorry laden with watermelons. During our faltering chats with the driver, he offered us some of his fruit. We went back to our villa with a car-boot-full, and they triggered my plan to cook the pepper and tomato dish from Paris. I made the melons into a soup with a *jamon* bone, which felt like the right kind of prelude to the stew. To my amazement the latter proved to be piquantly presentable, given I'd never cooked anything like it before. We dined lounging on pine fronds on the floor, like decadent Berbers, imagining we could see mirages of the North African coast from the veranda.

Later that year I bought a paperback copy of *Round the World in Eighty Dishes* which Penguin had published in 1962. It was my very first cookbook and I found it racy, exotic, more than a bit hippy. And there, in the section on Africa, was the dish I'd seen cooked outdoors in Paris, named as 'chakchouka'. Lesley Blanch had eaten it in Tunisia, sitting on the red earth in a troglodytes' cave and 'watching the women in their blue robes hung with amulets'. Her recipe, simple enough, was exactly what I'd witnessed, but was preceded by a cameo of Tunisian cafe society infused by her enthralment with oriental romance and painterly eye for detail: the Arab tea-drinkers gathering at dusk with their tame songbirds, 'often with a spray of jasmine held in their hand, or a rose tucked over one ear [and] beside them on the tables ... small cages made of porcupine quills'. This was heady stuff for a twenty-year-old with bohemian pretensions and not much experience of the world, and it kindled an enthusiasm for exploring, however vicariously, other places through their food.

Blanch wrote *Eighty Dishes* in New York in 1955. She was revelling in the success of her just-published tale of four women adventurers in the Middle East, *The Wilder Shores of Love*, and enjoying a brief spell of freedom from her moody husband, the novelist and diplomat Romain Gary, who was recuperating from one of his many spells of nervous debility in Berkshire. She'd originally intended it as a practical cookbook for children, but quickly became bored by the need for uncomplicated explanations and accurate recipes. What emerged instead was kind of sensuous travelogue presented through the medium of food, a guide to cooking as an expression of the indigenous imagination.

The very first recipe, for *pain perdu*, France's answer to bread and butter pudding, sets the tone and pattern for the rest of the book. Its story is set in the ramshackle property Lesley and Romain owned (and were forever restoring) at Roquebrune in Provence. They have unexpected guests one day, and are faced with an empty larder and what in most households would be an additional disaster – builders *in situ*. But it's these resourceful artisans who come to Lesley's rescue. They go off to procure a chicken, fetch veg from their allotment gardens. The foreman Marcel takes on responsibility for the pudding itself: 'I shall always remember him,' Lesley writes, 'far too large for the kitchen, covered in cement splashes, his huge paws delicately beating eggs and measuring sugar.' Here are all the ingredients of the food scenarios that follow: an intensely evoked and exotic setting (stage design was one of Lesley's polymathic skills), a cast of gifted amateurs, and a conviction that what gave food its character was not principally *terroir*, or the quality and provenance of ingredients, or any of the other imperatives of modern foodie-ism, but local temperament. The end products of her extravagant and witty tone poems are often rather humdrum. What is important, she stresses, is not what is cooked but *how*. Cuisine should be a passionate distillation of the *genius loci*.

There are more European encounters. 'Belgian cucumber' (sliced and poached in yogurt) was offered to her in a Flanders restaurant where the chef was hurling cleavers through the serving hatch at one of the waitresses. She is invited backstage at a puppet theatre in Sicily, and given the almondy sweet, frangipane, to nibble while the puppets were patched up. 'Papa Guiseppi [*sic*] was mending a tin sword, bent in the violence of a battle. His daughter, Tommasina, was heating some curling irons, to crimp the long golden wig worn by the Queen' – and the whole troupe sing snippets from the local light opera *Cavalleria Rusticana*. In Portugal she eats from lemon shells stuffed with mashed sardines and eggs round a bonfire on a beach, with the usual cast of characterful extras and a soundtrack of fado, 'those beautiful songs that recall the softer sort of Gipsy music, with an almost Oriental or Turkish cadence'.

It's when she taps directly into that Eastern cadence that the book has its best moments. Lesley had been enthralled by Slavic and Middle Eastern culture since she was a child, and had an enviable freedom to follow her enthusiasm, accompanying her husband on diplomatic postings in the Balkans and filing journalistic assignments for *Vogue* and *Harper's Bazaar*. And it's in the anecdotes she tells from Bulgaria, Romania, Russia, Turkey, Lebanon and Syria that you most clearly appreciate that this isn't a cookbook based, as many are, on classic restaurant food. Its dishes are streetwise, sassy, intimate. Lesley was irresistibly vivacious, oblivious to class and inquisitive to the point of nosiness. She seemed able to charm her way into any situation where interesting food was happening, from an Arab nomad's tent to the kitchens at embassy banquets. She picks up recipes in bazaar kiosks, Gypsy encampments, sheikhs' palaces, shanty-town cafes. In the Caucasus she is thrilled by Georgian tribesmen who 'sometimes grill the meat on their swords; then throw a glass of brandy on it, so that it is brought to the table flaming. And sometimes they dance, brandishing the lighted swords, their tall black sheepskin Kolpaks cocked over on the side of their heads, as you can see in my drawing.' (Her brilliant vignettes are one of the book's bonuses.) This leads to a modest recipe for shashlik on skewers, since the sword dance is 'not the sort ... that can be done in the average kitchen'. The settings, always more prominent than the cooking 'plots', provide the same kind of contextual mood as background music.

In Jeddah in Saudi Arabia, which looks 'as if it is made of ivory, with cream-coloured carved and decorated houses', she penetrates the harem, and eventually picks up a recipe for apples stuffed with spiced chicken called *foudja djedad*, a very cream-coloured, carved and decorated dish, its mixing of meat and fruit then unfamiliar to most British cooks. The signature dish of her Balkan chapter is what she calls Bandit's Joy (she's addicted to eye-catching *noms de cuisine*). She learned this from the sister of an Albanian outlaw with whom she confesses she was once 'rather friendly'. Most of the account is taken up with a lip-licking description of his rakish character and outrageous appearance – 'a very handsome creature ... His head was shaved, but he had enormous black moustachios. He wore the white felt cap most Albanians wear, at least three embroidered jackets, a sash stuck with knives, tasselled trousers, and over all a huge shaggy sheepskin coat.' His 'joy' turns out to be nursery food – boiled potatoes with butter and honey: 'a very odd choice', Lesley comments, 'for such a violent man'. But then Lesley liked her men sweet and sour.

It's here you begin to wonder if a bit of mischievous make-believe has entered the text. Did Lesley really have all these romantic adventures, extract the culinary (and other) secrets of every sultan and servant girl she met? It isn't unfair to see her as a kind of Scheherazade, spinning elaborate tales of her gastronomic Arabian nights. This, after all, was her style in life and in her writing. And it was part of the aim of her book to insist that food was an imaginative and at times fantastical art. Her stories of the origins of dishes are full of myths and folk legends, of golems and lost children and outrageous discoveries: 'One classic early-nineteenth century Russian cookbook contains a recipe which opens on this flamboyant note: *Take the yolks of five hundred eggs ...* I wonder what happened to the whites.'

In some of the final sections, 'The Far East', 'The Pacific', 'South America' and 'The North', she has no choice but to go beyond her personal experience. She hadn't travelled in these regions by the time the book was originally published, and relied on friends' experiences and cribs from texts such as Countess Morphy's *Recipes of All Nations*. But even when she is out of her knowledge she makes magic with her food. I'm sceptical that she ever ventured far into Haiti, and whether there is really a local speciality called 'The Zombie's Secret'. But after some hokum about voodoo, she describes a dish of mashed avocado, banana and cream cheese, flavoured with strong coffee dreamed up by a zombie. It's a very agreeable concoction (I've made it) but smacks more of some faux-tropicana dessert by Nigella than an invention of the undead. I still relish the sauce for baked bananas she makes up herself, from a brew of rum, marmalade and brown sugar.

I think it was this skittishness and spirited sense of making do (both her own and her teachers') that made *Eighty Dishes* such a hit with us aspiring gastronauts in the sixties. Away from more functional food-making at home, we were doing our experimental cooking in cramped boat-galleys and disused barns, which added just a hint of Lesley's exotic kitchen sets. Elizabeth David's *Mediterranean Food* preceded Lesley's book by five years, but by comparison David's recipes (they covered many of the same dishes) seemed to me a touch pedestrian, devoid of intimate detail and any strong sense of local culture. When I came to write an offbeat cookbook myself many years later (*The Full English Cassoulet*, 2008) it was Lesley's gleeful delight in improvisation that was my inspiration.

For some of us *Eighty Dishes* was a first glimpse of the seductive delights of 'abroad', and maybe that is a charge that can be made against it, that it's a throwback to a naive and patronising orientalism, an exotic version of the pastoral. In it the peasant women are always dark-eyed, the men always singing while they work, the courtyards decked out with chromatic colour. The book is set in some of the poorest regions of the globe, but there is no hint of hunger or oppression. I suspect Lesley's counter would have been that it was precisely through inventive improvisation that people challenged deprivation and asserted their independence. In a note to a 1992 edition, written when she was eighty-seven, she gently parodies her own style with an imaginary postcard home: 'Muscat. Supper with pearl divers on their boat. Dashing lot. Shark stew and prickly fig jam for pud.' But then adds a poignant postscript, reminding her readers '*when* those recipes were collected ... Since then, all over the world ... what is described as progress [has] swept away tradition and local colour, replacing them with a remorseless unification or desolation.' She is too pessimistic. Indigenous food survives in new contexts, not least because of her spicy record of it.

The New Foraging

First published in the Guardian, *2010*

The omens this autumn have been auspicious: drought-baked earth, stressed roots, torrents of rain – the same conditions which brought about that *annus mirabilis* of wild fungi in 1976. The fruiting that season permanently dented our ancient suspicion about eating the Devil's Meats, to the extent that in my home town in the Chilterns entrepreneurial schoolkids were hawking bags of horse mushrooms door to door. Thirty years on the foragers will be out again. If you want a field guide to their curious and atavistic behaviour, read Gary Alan Fine's *Morel Tales*, an anthropological study of 'the culture of mushrooming' in the United States. Its stories are edgy, almost surreal at times, but their combination of urban sophistication and backwoods yearning is recognisable in our own foraging culture. There is the trace of gung-ho competitiveness (whose puffball is biggest). There is the ecstasy of discovery: 'Suddenly it is there in the shadows. A single, exquisite morel … stands by itself, boldly etched against the edge of the orchard. Awestruck at first, I am afraid to remove it. Perhaps it is the last morel in the world.' A Minnesota man talks about the quality of 'gatheredness' that makes wild foods taste different from shop-bought ones. And a forager mistakes a child's plastic ball for a puffball, and is still chuffed by his hunter-gatherer's sharpness: 'It looked *exactly* like the real thing.'

In Britain, foraging in a more general sense is going through one of its more active periods, which have tended to spring up in times of nostalgia for lost rural values. As I write, at least two television series are in preparation. Local authorities are laying on guided forays in country parks. Star restaurants are offering their customers the chance to go feral with *bonnes bouches* like nettle froth and rowan sorbet. Lower-order gastropubs tuck dandelion leaves into their salads, and boast of 'wild mushrooms' in their stews. And this summer, supplies of citric acid (kept under the counter in most chemists because it can be used to cut heroin) ran dry because of a national craze for making elderflower cordial, in which it's an essential ingredient. Most telling maybe has been the change in the fortunes of that seashore delicacy marsh samphire. In the sixties it was an arcane, liminal food, a 'poor man's asparagus', gathered and eaten strictly locally, chiefly along the east coast. In 1981 it was served at Charles and Diana's wedding breakfast, gathered fresh from the Crown's own marshes at Sandringham. Now (imported out of season from Israel) it's a widespread garnish for restaurant fish; a seaside holiday souvenir, sold by the bag to those who don't want to get their own legs mud-plastered; and, along the north Norfolk coast, an occasional bar-top snack, served lightly vinegared in bowls next to the crisps and pickled onions.

All of which raises the question, why? Why should twenty-first-century eaters-out, with easy access to most of the taste sensations on the planet, choose to browse about like Palaeolithics or primates? They're opting for the most part for *in*convenience food, for bramble-scrabbling, mud-larking, tree-climbing. For the uncomfortable business of peeling horseradish and de-husking chestnuts. For the sour, the insect-ridden, and for the historic duty of eating the frankly rank ground elder, just because it was brought here as 'a pot herb' two millennia ago. A far cry from the duty that Edward

Hyams, doyen of the domestication of plants, spelt out – 'to leave the fruits of the earth finer than he found therm' (*Strawberry Cultivation*, 1943).

I made my own contribution to the literature of foraging back in the early seventies, with *Food for Free*. I'd been spending long spells in a rented cottage on the Norfolk coast, birdwatching and beach-rambling, and discovered that a tradition of garnering edibles, especially from the shoreline, was still flourishing. I'd been casting about for a subject for my first full-length book, and this seemed a natural. At the back of my mind was a sense that the time might just be right for such a venture. There was a growing public worry that our food was becoming toxic and bland, parallel with a concern that we were 'losing touch with nature'. The rationale I spelt out in the introduction was accordingly earnest, not to say pious. I lamented our remoteness from the origins of our food, our conservatism with ingredients, our wastefulness. I reminisced about a vanishing local knowledge of wildings, the root source of all our cultivated vegetables. And I talked excitedly about a lost world of exotic scent and flavour – the apricot savour of chanterelles, the avocado texture of hawthorn berries – and about how the hunt sharpened all your senses ('search *inside* hazel bushes for the nuts, as well as working round them, and scan them with the sun behind you from the inside, so you can glimpse the nuts against the sky'). But a snapshot from the seventies reveals another agenda. I'm sitting cross-legged on the lawn in a kaftan, and cradling, with something of a snigger on my face, an immense puffball. This wasn't tapping into some ancient rural heritage, or an exercise in survivalism. It was counterculture stuff, hedge-wise, irreverent, two fingers waved as much at domesticity as food domestication. Others might have recognised similar posturing in my research routine on the Norfolk coast. I'd prowl about like a modern hunter-gatherer, binoculars round my neck and a capacious string bag over my shoulder. When I got back to my cottage, I tried to do something inventive with the weed of the day, and washed it down with a bottle from the Wine Society. Survivalist I was not.

While I was writing I spent as much time in libraries as in the field. I raked through the diarist John Evelyn's waspish vegetarian tract, *Acetaria, A Discourse on Sallets* (1699), through learned papers on the stomach contents of mummified neolithic corpses, through Eric Linklater's novel *Poet's Pub*, searching for a half-remembered medieval recipe for a roast crabapple and beer winter cup. I was fascinated by the contexts in which this adventurous eating was being explored, as well as its improbable ingredients. My rule of thumb was that any plant which hadn't been specifically ruled out in the literature was worth a try. This resulted in some nasty taste experiences (raw sloes were the worst) but never, thank goodness, outright poisoning.

Modern foraging has always had more roots in zeitgeist than in self-sufficiency or hunger. It has chiefly been a leisure pursuit of the educated and well off, and not of those who genuinely need a cheap meal (though in an odd expression of the fantasy of living off the wild, sales of a cheap pocket edition of *Food for Free* soared immediately after the financial crash in 2008). As if to confirm this, many of its roots have been literary. One of the first accounts of local edibles, from a writer who'd tried them himself, came from my own foraging territory on the Norfolk coast. Lilly Wigg's *Esculent Plants* (c.1790) is a handwritten MS in the British Museum. The author was from Great Yarmouth, and worked, in turn, as a shoemaker, schoolmaster and bank clerk, setting out an early sketch of the middle-class artisan profile of the modern forager. Much of the wild food writing of the nineteenth century was in an enthusiastically participant, folklore collector's mould. The fungus foray was pioneered by the Woolhope Field Club in Herefordshire, a band of leisure-rich amateurs who were on the cutting edge of Victorian natural history. Their *Proceedings* for the autumn of 1869 describes a fungus-gathering expedition by thirty-five of their members (nine of them vicars). They ranged around the local landscape by carriage, stopping off at likely hunting grounds, measuring fairy rings, and gathering an extraordinary hoard of wild mushrooms: milk-caps, ceps, chanterelles, witch's butter, hedgehog fungi. The day ended in the Green Dragon in Hereford, with the exhibits strewn out on the pub tables, and a late lunch of the day's best trophies: shaggy parasols on toast, fried giant puffballs, and fairy-ring champignons in white sauce. The puffballs especially were voted a great success, as was the day itself.

In the 1930s, a romantic, pastoral note entered wild food writing. Florence White (founder of 'the English Folk Cookery Association') described her collection of recipes as an 'intriguing' literary hobby that 'proved full of unexpected interests. It brought the country into the town house or flat.' In her *Flowers as Food* (1934) she encouraged her readers to experiment by making 'small quantities of flower syrups, vinegars, herb-teas, wines, confectionery etc'. Dorothy Hartley, a redoubtable writer and artist, toured the countryside by bike between the wars, collecting traditional recipes. Her quirky masterpiece, *Food in England*, wasn't published till 1954, but is stuffed with wild snippets: the Kentish hop-pickers way of cooking their trimmings, like a thin asparagus; an ethereal blackberry junket made simply by leaving the strained juice in a warm room; and, anticipating that breathless American paean to morels, a eulogy to chanterelles, 'sometimes clustered so close that they look like a torn golden shawl dropped down among the dead leaves'.

The underlying patriotic tone of these books found its moment during the privations of World War II, one period when foraging did become functional. Vicomte de Mauduit's splendidly titled broadside, *They Can't Ration These*, was joined by the Ministry of Food's own pamphlet, *Hedgerow Harvest*. This moved the Home Front out into the wild, with recipes for rosehip syrup – given to children as a source of vitamic C – and sloe and marrow jam: 'if possible crack some of the stones and add to the preserve before boiling to give a nutty flavour'. This was an epicurean touch, but prefaced by the first strictures about picking etiquette. 'None of this harvest should be wasted, but be exceedingly careful how you gather it in ... don't injure the bushes or trees. When you pick mushrooms, cut the stalks neatly with a knife, leaving the roots [*sic*] in the ground.'

When the foraging habit resurfaced in the 1970s, it was among a mixed constituency of hippies, supermarket refugees, organic foodies and nostalgic countryphiles, and quite soon two separate, though far from exclusive, factions emerged: those whose interest in wild food was principally about the food, and those who were enthralled by the wild. Hugh Fearnley-Whittingstall's books and TV programmes (e.g. *A Cook on the Wild Side*, 1997) have made him the guru of the food faction. But the absorption of wild food by conventional – and commercial – food cultures was already under way. Elderflower gathering – and eventually growing the bushes in 'elder orchards' – became big business in the Cotswolds. So did the garnering of Argyllshire heather tops for the honey-scented ale sold as 'leann fraoch' from Bruce Williams's Glasgow brewery. And in tune with the increasing coachloads of French mushroomers scouring the Sussex woods, the United Kingdom became a net exporter of wild fungi.

In the 1980s I was recruited as an adviser by a team of restaurateurs and publicists planning to start an all-wild-food restaurant in London. The site chosen was Covent Garden, which was ideal in terms of its likely food-savvy clientele, but seriously disadvantaged by its distance from the raw materials. None of us had thought through what obtaining even an abundant product like nettles meant in terms of gathering enough for forty servings, and transporting it quickly from the countryside to central London. The restaurant worked for a while, but only by compromising its all-wild ambitions. I should have remembered the herculean efforts needed to mount a wild food buffet for the launch of *Food for Free* in the autumn of 1972. Friends ferried in mushrooms and hazelnuts and hedgerow jams from across the Home Counties to the publisher Collins' canteen in Mayfair. One drove most of the passengers out of an Underground carriage with her armful of liquorice-scented wild fennel. Collins' canteen staff were so alarmed by the produce we were bringing in that they locked up their refrigerators.

These days most smart restaurants incorporate some wild foods in their menus by employing a new class of professional foragers and using the ingredients in small quantities. The menu at Simon Rogan's *L'Enclume* in Cumbria reads like a collection of food haiku, spiked with Cumbria's indigenous sproutings: 'Poached brill, English mace, wild tree spinach, pickle from then mixed with a bit of now .... Bottled aromas, sweet cicely, passion pipetted.' But the poetry – and honest usefulness – of vernacular weeds have flowered in the humblest of spots. In the small Suffolk village of Milden, named after that Neolithic staple vegetable fat-hen (*melde* in Old English), the council has put up a cast-iron statue commemorating their name plant, which the local farmers are forever trying to eradicate from their sugar-beet fields. In the United States, meanwhile, it is the wild tendency that has flourished, with practical but lyrical guides like Euell Gibbons's *Stalking the Wild Asparagus* (1962), and Alix Kates Shulman's feminist forager's memoir, *Drinking the Rain* (1995), in which she drops out to a shack on the Maine coast and comes to terms with middle age by feeding herself from the shoreline.

For myself, I dithered between the two schools. The nutritional tendency began to seem too dull and correct, and I was terminally alienated by a meeting with a distinguished portrait photographer about ten years after *Food for Free* was published. 'Ah yes,' he pronounced, eyeing me as if I were an exotic species of root vegetable, 'you're the man that eats weeds. What an interestingly *earthy* face.' But I hadn't the temperament to be a full-blooded back-to-nature survivalist either. Instead I flitted about the hedgerows in my native Chilterns, in the guise of a hopefully down-to-earth aesthete, searching for wilding apples. Hunting for these happenstance fruits, sprung from thrown-away cores and bird droppings, seemed to catch all that was exhilarating about foraging: a sharpness of taste, and of spirit; an echo of the vast genetic inheritance of cultivated edibles; a sense of possibility, of things that happened beyond our ken. I can still recall the trees along my apple elysian way, a stretch of ancient green-lane strung along the edge of a valley: one sniffed out from fifty yards away – lemon-yellow fruit, scent like quince, too acid and hard to eat raw, but spectacular with roast meat; another with the bitter-sweet effervescence of sherbet; a third with long pear-shaped apples, and a warm smoky flavour behind the sharpness, as if they had already been baked.

Henry Thoreau loved his wildings, and in his essay 'Wild Apples' he writes a wonderful litany of cod botanical subspecies and datelined varieties: 'the apple that grows in dells in the woods (*Malus sylvestivallis*) ... the Partridge Apple; the Truants's Apple (*Malus cessatoris*), which no boy will ever go by without knocking off some ... the December eating; the Frozen-thawed (*Malus gelato-soluta*), good only in that state ... the Hedge Apple; the Slug Apple; the Railroad Apple ...' Thoreau understood the quality of 'gatheredness', that the circumstances of finding and picking wild food somehow infuse its whole savour.

Foraging came of age in this country in 2004 when the Oxford Food Symposium – that annual gathering of erudite foodies – chose 'Wild Food' as its theme. It was a chance for the wild and the food factions to hear each other's points of view. There were learned papers on 'Wild Food in the Talmud', and on 'Fake hare ... and Fake leg of Venison: Recipes for Ersatz game in 19th century German Cookery Books'. There was a tasting of 'The Feral Oils of Australia', a remarkable collection from trees which had self-sown in streets and waste ground in the island's hospitable climate. Caroline Conran gave a key address on her work on the south-west French tradition of *la cueillette*, the seasonal gathering of wild mushrooms, berries and greens. This is still a useful local bounty, but no one would pretend it's either essential for survival or commercially important any more. Rather, Conran suggested, in an argument that has resonance with foraging over here, *la cueillette* is a ritual, a reassertion of communal identity and territorial rights. The people of southern France had their independence hammered first by the *ancien régime*, and then by the Revolution, which disapproved of the idea of common land and sold much of it off for state forestry. The modern *cueillette* is a rehearsal of ancient rights, a snub to the idea of nature as private property or domesticated pet, a celebration of belonging to one's *pays*.

Many speakers, if sometimes uneasily, echoed the ritual value of foraging. It puts you in touch with your roots, they argued, with the very notion of food as a product of natural processes. It sharpens your sense of the seasons, your reading of the landscape. Thoreau, again, was much quoted: 'The bitter-sweet of a white-oak acorn which you nibble in a bleak November walk over the tawny earth is more to me than a slice of imported pine-apple.'

Thoreau's unfinished manuscript *Wild Fruits*, from which this sentence comes, wasn't published until 2000, but has become a seminal text for New Foragers. It's a flamboyant celebration of tanginess, of the buffers that agriculture and commerce put between human eaters and the staffs of life. In the description of 'Going a' Graping', he relishes the difficulties of harvesting wild grapes, the scrambling and crawling and nosing-out, and stresses that these are as much part of the harvest as the grapes themselves: 'What is a whole binful that have been plucked to that solitary cluster left dangling inaccessible from some birch far away over the stream in the September air, with all its bloom and freshness.' The otherness of the wild. This is the paradox which is inherent in Thoreau's view of nature, and one which all foragers have to grapple with: nature wild and free on the bush, or captured and rendered down into some fancy *jus*.

Having a mite of responsibility for the revival of UK foraging, I said my piece at Oxford too – rather hesitantly, because I've got reservations about the whole business now. Such wholehearted, Thoreauesque relishing of armfuls of the wild might have been acceptable in the wildernesses of nineteenth-century America, and even now in the empty expanses of south-west France. But could a built-up, over-farmed and over-populated country like Britain really sustain foraging as more than a minority habit?

A finders-keepers morality didn't seem to me adequate any more. Didn't we need a kind of foraging ethic, which regulated our leisure whims in keeping with the needs of other organisms in the ecosystem?

As it happens, I've become more of a wayside nibbler than a forager these days, probably out of laziness as much as ecological worthiness. I like seren-dipitous findings, small wayside treats. Lately I've been pleased by having to use fennel flowers in a sauce, from a plant whose green leaves had been shrivelled in the heat; by finding a bough from a roadside damson bush that had been flailed while it was still in fruit – and experiencing the improbable taste of sun-dried English prunes. And in my rucksack, waiting for a moment of courage, are a handful of thistle galls, wasp-grub-induced swellings on the stem, like miniature kohlrabi. Can I go down this path? Can food?

I've found a phrase to catch that feeling in the fruit aesthete Edward Bunyard's book, *The Anatomy of Dessert* (1933). Talking of evening meanders through his gooseberry patch, he describes the pleasures of 'ambulant consumption'. 'The freedom of the bush should be given to all visitors.' The freedom of the bush. What a liberty!

But then, a quite different way at looking at self-discipline, at an etiquette for consumption. Down the lane to our common, there's been a vegetable roadkill. Wind-thrown cherry plums, yellow and scarlet like a painted doll's cheeks, are scattered all over the road. They're rare fruiters here, and almost without thinking, I'm crouched on the tarmac amid the puddles and trash, stuffing them into my pockets. And it occurs to me that this is how many of the planet's citizens, of all species, find their food. Opportunistically. Scavenging the surplus. Making do. Somewhere between this variant on the old idea of gleaning and 'ambulant consumption' – between working the margins and working the imagination – might lie the outline of a code for planetary foragers. And, just as foraging is a metaphor for a larger connected relationship with nature, mightn't such an ethic have the rough shape for a model of responsible consumption? So just a side serving, please.

The Unofficial Countryside

First published in the Sunday Telegraph, *2012*

When I was in my mid-twenties, I landed a job in publishing, in London. Actually it was in Middlesex, barely even Greater London, and out in the hinterland of peeling warehouses and run-down gravel pits north of Heathrow Airport. But it was a hot editorial job at Penguin Books' new education division, and I felt as cranked up as if I'd been beamed from my beechwood homeland direct to the pulsing heart of Bloomsbury. I'd spent the previous two years working as a lecturer in social studies at a College of Further Education in Hemel Hempstead New Town, and the chance to get into the book world through the back entrance, so to speak, was irresistible. Penguin Education had been set up by Sir Allen Lane to, in his words, 'carry the values and spirit of Pelicans [Penguin's non-fiction imprint] into the classroom', so there was a real social and cultural excitement in prospect, too.

But landscape, as so often, intruded its ambivalent tendrils into the mixture. I was still living in Berkhamsted, without a car, and I did the hour-long commute from the Chilterns to the offices at West Drayton by coach. I felt as if I were watching a road movie through the windows. The middle part of the journey passed through the valley of the River Colne, long since scalloped into pits and lakes of all kinds, and threaded with aspirational new highways. I had glimpses of fugitive water and resplendent waterbirds, glinting behind thickets of wayside scrub. Everywhere immense earth-movers scrunched like glaciers through prehistoric beds of flint and gravel. On a new roundabout near Uxbridge they'd piled up an artificial cliff-face of sand, and during that first summer a colony of opportunist sand martins raised their families in the shifting bank, darting nonchalantly between the JCBs and the traffic.

It was an even more fantastical scene at work. Our outpost of the Penguin empire was shaped like a slice of cake, and wedged between a canal and a main road. When I looked through one window of my office, I gazed down on a flotsam-strewn canal edged by plants I'd never consciously noticed in my life before, from all quarters of the globe: Indian balsam, Japanese knotweed, Canadian fleabane. From the other window I could watch kestrels hovering above the rush-hour traffic on the canal bridge. I tried to imagine the complexity of the images on their retinas: towpath jungle, pub garden, concrete and metal stream, all in different focal planes.

I took to exploring this hybrid landscape in my lunch hours, and found its rowdy, cosmopolitan luxuriance played out on a grand scale. I wandered across derelict rubbish tips, where bottle collectors scraped away for Victorian medicine phials and the products of 'Botanic Breweries'. I was still learning plants on the hoof, and was amazed at how cohorts of ambitious garden escapes had found their way here. Along the banks of the canals a mixture of rampant Asian and Mediterranean shrubs – buddleia, bladdersenna, Russian vine – was beginning to form a kind of transcontinental wildwood never seen on the planet before. I discovered a single plant of the magnificently malevolent thornapple, whose spiny fruits were once used as an anaesthetic, and wondered if it had sprung from the relics of an old herb garden or just from a stray seed in a pot of South American dahlias. And, by a service road to a scrapyard, there was a twelve-foot-tall specimen of giant hogweed, from the Caucasus originally but probably carried here from a relict Edwardian garden. It was surrounded by a cordon of blue safety tape, as if its bristling, rash-raising stems were already a crime scene. There were metropolitan birds too, terns wafting over pits where the dredgers were still pulling up buckets of gravel, and great crested grebes nesting on floating car tyres. I was bedazzled by all this incongruous luxuriance. This wasn't what was supposed to happen. Inspired by the architectural critic Ian Nairn's *Outrage*, our generation imagined that the brave new world was predicated on sweeping away 'clutter and sprawl'. I'd shared the deeply anthropomorphic assumption that the remnants of industrial dereliction were as inimical to nature as to humans. But here nature seemed to have a different agenda, an insistence on a postmodern coexistence with the city, even a hint of triumphalism. Its mana had no connection whatever with the cosy tranquillity (or so it had seemed) of the landscapes of Deep England, real and imagined, that I'd grown up with and now felt that I must have misread. The West Drayton wastelands presented another face of nature, as canny, adaptive, at times positively bolshie. In the charged atmosphere of the late sixties it seemed almost insurrectionary. In Paris the students were manning the barricades. In our office we were generating collections of vernacular poetry and student-centred science that would blow the conventional image of the school textbook apart. And outside the office a wild, inspirational, multicultural jungle was also defying traditional boundaries.

Its resilience, so at odds with the orthodox view of nature as vulnerable dependant, began to affect me. I was already well into my book on edible wild plants, *Food for Free*, and thinking about what might follow it. I reckoned what I had casually nicknamed the 'unofficial countryside' might make a viable subject, and began to explore it more methodically. I went on carefully planned urban safaris, not just near the office, but in other parts of London and even in the more built-up regions of my home town. One summer I did a two-day walk round the whole of inner London, beginning in the west in the foxy barrows of Kensal Green Cemetery and passing, at the eastern extreme, through the phantasmagorical landscapes of Stratford Marsh. This remains one of the most thrillingly scary environments I have been in. I slithered along flooded canal towpaths with my back pressed against the old brickwork of tunnels, passed tangled drifts of blood-red-tinted danewort, stinking of butchers' shops. On the far side of the cut, the metallic calls of black redstarts were ringing from a breaker's yard full of red phone boxes. Turning west again, I passed along Cable Street, where the East End had battled Oswald Mosley's blackshirts in the 1930s. I passed patches of so far undeveloped wasteland which had turned into buddleia forests. House martins (still common then) were flying in from their nest sites on Thameside properties in Wapping and the Shadwell Basin, and flitting for insects above the bushes. I saw a kestrel glide out of the window of a derelict Victorian school (with separate 'Boys' and 'Girls' entrances.)

I began to learn something about the streetwise ways in which nature adapts to our occupation (and disruption) of its ancestral territories. I saw city pigeons embarking and disembarking from Tube trains, en route to new feeding stations. At Perry Oaks sewage farm, just a couple of miles from my Penguin office and perched at the end of Heathrow's main runway, I watched migrant waders – greenshank, little stints, sandpipers – using the settling beds as a substitute mudflat. I tried to identify the various species of Asian knotweed that had escaped from herbaceous borders to ramp on all kinds of waste ground. In Buckingham Palace gardens with a film crew I met a venerable entomologist (the splendidly named Baron de Worms) who was manning a moth trap, and he told me the story of a rare African clothes moth that had turned up at his mercury-vapour lamp after a Commonwealth Prime Ministers' garden party.

This was cheekily funny, but all urban nature is a kind of comedy in the Shakespearean sense. It's about transformations, disarray, mistaken identities (and addresses) and things being generally where they shouldn't. It challenges our expectations and neat categories of the world.

These formal expeditions weren't, as it turned out, the way to tap into the spirit of urban nature. You can't reduce raucous green street life to the orderliness of the rural, make it into an enclosure map. There are charts of it, but they exist in other dimensions, and are constantly on the move.

In the summer after the Blitz, there was weed-storm across London. The bomb sites were covered with a purple surf of rosebay willowherb (unfamiliar then but quickly tagged 'bombweed'). Damp-loving ferns carpeted the wrecked shell of St James's in Piccadilly, drenched as it had been by the Auxiliary Fire Service's hoses. Medieval apothecaries' herbs sprang up from the newly sunlit cellars of the old merchants' quarter in the City. Londoners didn't know what to make of this invasion, whether to see it as a benediction, a sign of life returning to the ruins, or an insult added to the very obvious injury. A few thought the seeds had come in with the German bombs. But they were nothing new. In 1877 a well was sunk at a brewery in the West End, not far from what would become the epicentre of the bomb damage seventy years later. Quite close to the surface archaeologists found the fossil remnants of plants that had flourished in the Thames basin 20,000 years ago, when what would become the city was a rock-strewn tundra. They were the same species – bracken, horsetail, chickweed, rosebay, dock, buttercup – that were to reappear with a new and mystifying abundance in the 1940s.

Urban plant life – indeed all urban wildlife – is the familiar of disturbance. The organisms that evolved to cope with volcanoes and glaciers and melt-floods adapt well to wars and the creep of new roads. Their map isn't some ground plan of fixed locations, but a chart of rhythms – syncopated, opportunist beats, episodic ebbs and flows – as buildings rise and are knocked down, new communication routes open, new settlers arrive. It might, if one could pin it down on a screen, look like a map of the sands in an intertidal zone, shifting and ill-defined.

This fleet-footedness means that urban wildlife is remarkably tenacious. It survives not by staying put but by moving on. The current assault on so-called 'brownfield sites' – a misnomer for places which blaze in brilliant technicolour by comparison to the arable countryside – means the loss of the last truly wild spaces in cities. But by and large urban wildlife simply regroups and finds the next upheaval. Almost anywhere in Britain now you can chance on these ephemeral flourishes. There have been short-lived colonies of piranha in Essex, where aquarium owners have found that the appeal of these piscine Rottweilers fades fast, and simply dumped them in the nearest pit. Roosting flocks of rose-ringed parakeets, sometimes thousands strong, now transform parts of south London into passable imitations of dusk in India. For some residents, their mass screeching sours the spectacle of waves of lime-green and carmine-pink parrots flying above the Victorian terraces. But by a strange closing of circles, I saw a flock flash through the grey shrubberies of Kensal Green Cemetery during the funeral of the one-time head of Penguin Education, Charles Clark, and it seemed like a benediction.

Flamboyance and pushiness may have given a bad name to other species from beyond our shores, but no one has anything but affection for the fig woods that have become established along the banks of the River Don in Sheffield, where the great steelworks once stood. It's a remarkable story. In the 1920s, when the industry was flourishing, the river water was used as a coolant and the Don ran at a constant temperature of 20°C – hot enough for fig seeds washed into the river from sewage outfalls to germinate and eventually grow into trees. They are now thirty feet tall, have Tree Preservation Orders on them and, one local ecologist remarked, 'are as much a part of Sheffield's industrial heritage as Bessemer converters and steam hammers'.

But sometimes it is fleetingness itself that gives a charge to encounters with urban nature. I was once in a taxi in the Strand on a boiling late-summer day when a huge hawker dragonfly flew in the open window, hovered briefly, considered its situation, and then flew out again. I could not imagine what it was doing there in the leafless heart of central London, and pondered the possibility that ancient flyways were still etched on the city's territory, surviving all the transformations that had happened to the ground itself.

When I moved from the Chilterns to Norfolk in 2002, my metropolitan centre became Norwich. A different kind of city, but still with those ubiquitous urban eco-rhythms. 'Norwich, a Fine City' the boundary boards read. Like London it was badly bombed in the Second World War, but much of the medieval centre, a labyrinth of cobbled streets and ancient churches, survived and provides other echoes, other layers in the palimpsest of the urban green. Keep your eyes up high, or down low, and there are weeds, plain and simple, new and old, everywhere. The artist Jacques Nimki has made a piece about them, a 'Norwich Florilegium', a set of postcards of his own favourite (albeit ephemeral) clumps. He commemorates 'Dindle' (aka nipplewort) on a ledge above Boots: 'If you look at the Google satellite map, just past Swan Lane you can see the shadow of the plant on the pavement.' On the remains of Carrow Abbey, near the Canaries' football ground, birthwort still flourishes, a relic of the nuns' physic garden (it was a labour-easer and abortifacient). Norwich's focal point is the Norman cathedral, with its astonishing 315-foot needle-thin spire. Its walls are cloaked with plants representing a thousand years of the city's history, including Oxford ragwort, whose seeds hitched a ride here – by rail – from the Oxford Botanic Garden in the mid nineteenth century. The cathedral has a botanic garden of a kind too, a recreated Benedictine physic garden whose new plantings are also escaping out onto the ancient walls.

And those intrigued by the curious fusion of natural and man-made that urban wildlife represents will be drawn to the cathedral's cohort of Green Men. Eight of these carved foliate heads, mouths and nostrils and ears brimming with oak and hawthorn leaves, grimace down from the roof of the cloisters. The Christian Church appropriated the iconic form of the Green Man, hoping that the rampant foliage sprouting out of (or into) his orifices would suggest the sinful temptations of unredeemed nature. But it is an older pagan symbol, a relic from early Eurasian cultures when humans were seen as part of the natural cycle. You can read the Green Man in either direction. The new leaves, miraculously transforming sunlight, feed our mouths. Or our mortal remains feed the growing plant.

Two years ago the Green Man's symbolic coupling of the human and the natural happened for real two hundred feet above the cloisters. A pair of peregrine falcons, the most charismatic of our birds of prey, began frequenting the cathedral, viewing the spire – as they are beginning to see tall buildings all over urban Britain – as a simulacrum of their natural cliff-face habitat. They feasted on street pigeons and gulls, sometimes hunting in the dark for night-flying migrants like woodcock. When it looked as if they might take up permanent residence, the Hawk and Owl Trust put up a nesting platform. The pair took to it, but the female bird was young and inexperienced and her one egg failed to hatch.

One year on the male is back with another, more mature partner. It's mid-April and four eggs have been laid. The Trust has set up a booth and telescopes in the cathedral close and they are in almost continuous use by enthralled visitors. I'm there about noon, and the male bird is cruising around at about two hundred feet, keeping an eye on things. Suddenly he begins spiralling up southwards. One of the wardens follows him with binoculars, and guides the rest of us. We track him gliding in and out of the cloud base. He's found two other birds of prey (or they have found him), a buzzard and sparrowhawk, also floating unnoticed high above the city. There are a few moments of irritable tag, but no fighting. And then, quite suddenly, the peregrine tucks in his wings, and stoops. In the perfect shape of a folded anchor the hunter falls through a thousand feet of air at a speed of over 150 mph, over the green where the city's *parkourists* are practising their leaps, and over the Castle Museum (where what is currently the nation's most news-worthy painting is hanging for the spring: Titian's portrait of the goddess of the hunt shows Diana, surprised naked by the peeking human Actaeon, whom she turns into a frightened deer in retaliation, to be mutilated by his own hounds). A minute or so later the hunting bird reappears carrying a kill, flying with a dead pigeon one-third of his own weight. The human watchers gaze in awe. He settles, momentarily exhausted, on the fifth crocket from the top of the spire (a favourite perch) and then begins to pluck. White feathers drift down onto the cathedral surrounds and faultless lawns of the close.

Up there, he is probably the highest living thing in the city. He is visible, perched heraldically, from shopping malls and high office windows. Once I was in the very top storey of a car park and glanced through my binoculars towards the cathedral. There he was on crocket number five looking for all the world like the archangel Gabriel. And it is the same from the falcons' point of view. A webcam above the nest platform shows the whole of Norwich spread out like a dominion beneath the brooding birds' gaze.

Many of the urban wildernesses I haunted in the seventies have vanished. Everywhere the strimmers of the tidy-minded and the planning instincts of the gardener are gaining ground. There are marinas by the West Drayton canals. The jungles of Stratford Marsh are buried under the Olympic Park, and in waste patches nearby I have found the indigenous weeds sprayed out to make way for commercial 'cornfield weed' mixes. The Olympic Delivery committees have plans for a large, ecologically engineered reserve as part of the Legacy – which is good news for biodiversity but, for me, no substitute for the upstart wilderness that was there before.

I know that spontaneous, 'self-willed' vegetation is not for everyone, and that I may have acquired a taste for botanical slumming. Sensitively land-scaped green space would probably be a more inclusive compromise. But wildness is not just a question of content, of this species or that, disposed as we think fit. It is also about a *process*, of natural regeneration and succession, which has its own redemptive qualities for those who witness it. In these terms the urban edgelands are some of the wildest places on our islands.

Plants in the Wrong Place?

First published in The Economist, *2011*

Are weeds a category of plant life or of human reflex? Are they more a cultural than a biological creation? We might, as house-proud gardeners or agribusinessmen, dream of a world without them, but we don't often pause to think why they are there, or what our planet might be like without them. Rather brown probably. Rather damaged and impoverished certainly. Take out all the wild plants which have been categorised as weeds and we wouldn't have the grasses which were developed into wheat and led to the birth of civilisation, or most of our cultivated vegetables, from carrots to garlic. We'd have no Velcro, inspired by the hooked fruits of burdock, and their obstinate clinginess to dog's fur. Gardens would have no sweet rockets or Shirley poppies or variegated ivies. At least half the world's medicinal substances, from gripe water to morphine, would never have been discovered. And gone would be the kids' *lingua franca* of daisy chains, dandelion clocks, Chinese-burn grasses.

I'm generalising of course. Plenty of 'weedy' plants have no such virtues, just as plenty of non-weeds (I'd nominate laurel and leylandii) are riddled with incipient loathsomeness. It all depends on your perspective, on what you mean by a weed. 'A plant in the wrong place' is the commonest and most useful definition. But this leads to a kind of moral relativity. We tramp the lanes for blackberry bushes, and rout them from the garden fruit plot. Poppies are sprayed out of wheat fields but patriotically revered every 11 November. One subset of entomologists adores buddleia, the butterfly bush (brought here as a garden shrub from China more than a century ago), as a nectar source for flying insects; another abhors the way it has formed a kind of urban forest on waste ground, shading out the habitats and forage plants of the larvae of these insects.

All weeds are trespassers of a kind, but across boundaries which are as often cultural as topographical. Their blacklists are consequently so subjective that the only really practical definition is that a weed is a plant which gets up somebody's nose (literally in the case of hay-feverish grasses). Plants get labelled as weeds if they're toxic, ugly, parasitic (i.e. 'immoral'), aggressively loutish (Japanese knotweed) or just limp-wristedly 'weedy' (chickweed). In the USA, the Department of Agriculture, trying to find some unifying principle behind its own pragmatic lists, admits that 'over 50 per cent of the entire US flora is made up of species that are considered undesirable by some segment of our society'.

The one principle with which all social groups, on both sides of the Atlantic, agree is that they are not complicit in the presence or survival of these invading organisms. Weeds are 'the others', literal or metaphorical aliens. We have no wish to know about why they seem to dog human activities so remorselessly, only in finding ways of getting rid of the bloody things. They're regarded as inexplicable and impertinent intruders, whose presence is quite unconnected with the way we live our lives. Ironically it's only those fundamentalist Christians who still regard them as the 'thorns and thistles' with which God punished human's bad behaviour in Eden who get it right. Weeds are 'our fault'. It's the trashing of our metaphorical Eden over the past 10,000 years that has given them the opportunity to move out of the wild and into what we regard as our personal space.

Although any kind of plant can become a weed just by gatecrashing, the majority are species which evolved to cope with the naturally disturbed areas of the earth: on volcanic scree, along tidelines, in gale-wrecked forests. As soon as we began our human project of opening up the earth, be it with a fork, a JCB or a napalm bomb, we created replicas of these turbulent zones, which their floral natives smartly took advantage of. Nettles (native in river flood plains) move into the similarly rich and shifty soils of human dumps and fertilised fields. Bindweed scrambles up fences as effortlessly as it does willow scrub in the wild. Mexican fleabane, introduced to the UK as a rockery plant from the dry hills of central America, has escaped to become natural-ised on walls from the elegant perimeter of Merton College, Oxford, to rough drystones in the moors. Weeds are our most successful cultivated plants.

I touched earlier (p. 94) on how the plants that colonised London's bomb sites in the Second World War were the same as those discovered in excavated layers from 20,000 years earlier. These Palaeolithic weeds were doing their best to green over another broken landscape, shattered by glaciers and herds of rootling mammoth. I find it oddly cheering that there should be a category of plants which undertake this essential repair role. We need to deal with weeds when they directly obstruct our human affairs. But that shouldn't stop us respecting their role as nature's catch crop, part of its fabled abhorrence of a vacuum, components of a kind of vegetable immune system which does its best to repel the forces of entropy and 'development' that create barren-ness. And to do this they must be smart, nimble, adaptive and mobile. Many ashes (ironically called the 'weed tree' by foresters when they invade their tidy plantations) may soon be dead from dieback. But the space they leave will be quickly filled by other 'weed trees' – sycamore, from Eastern Europe, maybe sumach from the USA – which will force us to rethink our attitude towards so-called 'invasive aliens'. To those who would truly prefer lifeless brown earth to these opportunist green settlers, I can only suggest they book tickets for the first passenger flight to Mars.

PI

First published in Mondial, *2017*

Some summers back I was stretched out on a patch of English heathland, just enjoying the sun. Sprays of gorse and broom blossom were hanging like garlands against the sky, scenting the air with coconut and vanilla. Out of the blue I was hit by an extra burst of scent that felt as if it had been wafted directly at me. A few minutes later it happened again, and I recalled noticing these rhythmic gusts with other kinds of flower – lilacs, lime blossom, viburnum. I wondered for a moment if it was a kind of olfactory illusion, something in me not the plant. Sweet violets appear to produce their perfume in bursts because of a substance called ionone, which briefly anaesthetises our scent buds. The flower continues to smell but we momentarily lose the ability to register it. Or did some plant species budget the energy-expensive business of producing scent molecules as insect come-ons, and emit them as concentrated puffs? Then I had a more outlandish thought. Was the gorse smelling *me*, alert to the fact that there was another living thing nearby, and responding to my presence?

It was a piece of sun-baked whimsy, but not because it postulated plants having an active relationship with their environments. I didn't know it at the time but many species can detect potential predators and pollinators in their vicinity and adjust their behaviour accordingly. Bindweed can 'smell' the growth hormones in the saliva of grazing cows and responds by growing faster. Some bat-pollinated flowers move their petals to focus the echolocation signals of the bats. We're surprised when we learn of vegetal capabilities like this because we assume that plants are entirely passive organisms. 'To vegetate' in our lexicon of metaphors is to be barely one stage above being dead. It's a curiously defensive attitude towards a class of organism that makes up 99 per cent of the biomass of the planet, and that fixes the sun's energy for the whole of the rest of life. Animals we could do without, but if plants were to suddenly vanish, humans would follow them into extinction in a matter of months. Yet we continue to demote them to the status of planetary furniture – useful, decorative objects, but not worthy of consideration as agents of their own lives.

This dismissive history goes back at least to Aristotle, who grudgingly admitted that plants were alive, but with a 'low-level soul' that enabled their reproduction but nothing else. Ironically, it was the extravagant claims about plants' 'souls' in the 1970s that set back serious study of their abilities by decades. Books such as the bestselling *The Secret Life of Plants*, insisted that plants could be lie-detectors, practise telepathy, send out waves of spirit healing towards us, and, famously, presented us with the image of the evangelical Dorothy Retallack playing Bach and Led Zeppelin to her geraniums (she was trying to prove the morally corrupting influence of heavy rock). The intention was to elevate the status of plants, but in attempting to prove they possessed human gifts, it only served to obscure their special characteristics.

What chiefly differentiates plants from animals is that they are immobile, rooted to the spot. They can't escape predators, or go in search of reproductive partners. By way of compensation they've evolved a modular structure, with no irreplaceable central organs. Most species can lose 90 per cent of their tissue and regenerate from any remaining fragment. They have developed ways of having sex at a distance, and sophisticated communication and signalling systems that involve some twenty different senses. How is this possible, without any obvious sense organs, or any receiving and processing organ like the brain?

In 2014 two Chilean botanists chanced on some previously unnoticed properties of the Patagonian vine, *Boquila trifoliata*. This is an unexceptional plant at first sight, a woody climber which can spiral high into the canopy of temperate forests in South America. But its leaves are able to mimic the colour, shape, size and orientation of its host trees. Habitat mimicry is well known in the plant world, especially in places where it pays to be inconspicuous. Southern African *Lithops* species make a good job of disguising themselves as stones in deserts where every item of moisture-bearing food is hunted out voraciously. But this is an evolved, unvarying mimicry. What is unique (so far) about *Boquila* is that it seems able to change its leaves at will, morphing into facsimiles even of host plants it has never met before.

How it does this trick isn't yet understood. It may be that the vine's leaves sense volatile chemicals emitted by the host and these trigger cellular changes. But this wouldn't explain the subtlety of the mimicry, and the fact that is also takes in the physical shape of the host's leaves. More likely is that *Boquila* leaves carry advanced photoreceptors, analogous to those on the undersides of some moth caterpillars, which prompt changes in the insects' body colour to match the surfaces they're crawling over.

Light-sensitive cells occur in their millions all over the living surfaces of plants. They track the sun, help leaves avoid shade, respond to changing day length. They are multi-purpose and individually disposable, and work together, without the need of a processing organ, to control the plant's growth and orientation in space.

The same is true of plants' 'ears'. All plants possess receptors sensitive to touch vibration. Root systems appear to be especially responsive to sound, low frequencies encouraging growth, high inhibiting it. Plants also – perhaps coincidentally – transmit noises when growing strongly or under stress. These seem to be generated by the fracturing of cell walls. Put sound production and sound sensitivity together and all kinds of possibilities open up. The clicks from drought-stressed pine trees and maize stalks switch on moisture-conservation changes in nearby plants. Climbing beans look as if they have hitched the two modalities together into a kind of echolocation. Time-lapse photography of the tendrils doesn't show an indecisive meandering, ceasing only when a support is found, but a purposeful growth directly towards the nearest pole.

The most important channel of communication between plants, and between plants and the animal world, is smell, involving an immense vocabulary of volatile chemicals. It's hard for us as humans to get a handle on this. We recognise smells consciously. We're aware of their sources, and of the memories they trigger. We have trouble in envisaging unsmellable scents, and find the idea of them bypassing our conscious control and causing reflex behaviour decidedly disturbing, though we know it happens, when, for example, cohabiting women's menstrual cycles begin to synchronise. It is harder still to imagine how plants, lacking a single sniffing organ like the nose and an interpretative brain, are able to meaningfully interpret vaporous chemical signals. Plants certainly possess clusters of specialised cells for generating scents, but the sniffing cells seem to be widely distributed over the plant, and no one is sure how they work, or how minute concentrations of volatiles are converted into electrochemical messages which change the plant's behaviour.

The major role for floral scent is attracting insects. Flowers recruit pollinators with a variety of lures – aromas, ultraviolet colour trails, electrostatic charges and nectarous rewards. But for pollination to occur it is vital that insects forage principally among the same species, otherwise any pollen gathered is headed for a dead end. So flowers lace their nectar with traces of stimulant chemicals so that insects associate a particular combination of sight and sense cues with a pleasurable hit (caffeine is a frequent lacing of nectar). Many plant species have co-evolved with specific pollinators, harmonising their anatomies and aromatic messaging. The Madagascan Star of Bethlehem orchid has a twelve-inch-deep flower, whose nectar can only be reached by a moth with a twelve-inch-long tongue. Charles Darwin predicted such an insect must exist in 1862, forty years before it was discovered. It was named *praedicta* in his honour.

Orchids are the masters of scent communication. They are the largest family of plants on earth, and the only group to regularly mimic insect sex pheromones to attract them to flowers. The European bee orchids (*Ophrys* species) have blooms which roughly resemble wasps and bees, to encourage insects to 'pseudocopulate' with them, and hopefully pick up pollen in the process. But it is the allemone, the synthetic pheromone mimic, which draws them to the flower in the first place, often over long distances. Mexican bucket orchids take this process to several more stages of refinement. They exude a perfume which is irresistible to male euglossine bees, not because it resembles that of female bees, but because it mimics the pheromone produced naturally by males during courtship. Struggling to collect this potent wax, some of the excited males fall into a bucket-like structure in the flower, which is filled with liquid. There is only one escape route, via a narrow and convoluted passageway in which the pollen masses hang. After many hours (some orchid species have biological timers to control the entrapment) the sodden bee emerges, only to be attracted by another bucket orchid's flower, and submit again to the insect equivalent of Theseus's trials in the Minotaur's maze. Eventually the pollen sacs are transferred and the flower is fertilised. The male bee emerges, and once dried out is able to use the aphrodisiac wax attached to his legs in his elaborate display flight.

What we might call one-stage-removed scent communication occurs in other ways. Willows and oaks, for example, produce an airborne pheromone when they are under siege by leaf-eating insects. The original purpose of this was as a fast-track method of alerting the outer leaves that the plant is under attack, so that they step up their production of bitter defensive chemicals like tannin. But more distant plants – even other species – have learned to eavesdrop on this in-house signalling, and raise their tannin production too. I like the idea of other trees hacking inter-familial messages. Insects have learned this trick, too. Predatory mites and wasps, ladybirds and even real birds can read the trees' message on the wind that there is a concentration of insect food about. Mopane trees in Africa behave similarly, though in this case the predators are elephants, which are bright enough to argue back. The browsed mopanes' airborne warnings are blown downwind, so the elephants move *upwind* through the trees, taking just a few sprays from each, and hoping to outrun the vaporous messaging.

Underground, the chemical chatter is just as busy and even more social. It's long been known that the root systems of most land plants – trees especially – grow in intimate and mutually beneficial entanglements with what are called mycorrhizal fungi. The fungi, which have no chlorophyll, take the sugars they cannot produce themselves from the tree's roots. The trees in turn absorb minerals and other nutrients from the fungal network. What's only recently been discovered is that, in a forest for example, these networks can link trees of different species, forming not just a feeding cooperative but a vast signalling network conveying information as well as food, which some forest ecologists are now calling a 'wood wide web'. Like the Internet, it's reciprocal and non-hierarchical, with no single organism in the driving seat. Evergreens help top up the sugar reserves of the deciduous species in winter, and vice versa in summer. The fungus doubtless acts so that all its partners prosper.

This is all very smart, but much of this behaviour can be explained by the long processes of evolution, and old-school botanists are reluctant to attach the word intelligence to it. But learning new behaviour in your own lifetime, storing memories, problem solving – commonly accepted definitions of intelligent behaviour – that would be a different matter. The sensitive plant (*Mimosa pudica*) has been a source of puzzlement for centuries. As house-plant enthusiasts know, the merest touch of the feathery leaves causes them to fold inwards, starting an orderly wave of closure that ripples up the leaf stem. How and why it should do so greatly exercised the minds of eighteenth-century thinkers, more amenable to the ideas of a 'vital force' in plants than their modern counterparts. George Romney's 1789 etching of 'Emma Hamilton in an attitude towards a mimosa plant, causing it to demonstrate sensibility' shows Admiral Nelson's mistress, one of the most erotically charged celebrities of her time. One contemporary pronounced, 'that these plants *live*, will be granted, but I suspect that they likewise *feel*'. Philip Miller, author of *The Gardeners Dictionary* (1731), was more down to earth, and his summary of the various mechanical and hydraulic theories about mimosa's trick reads like a plumber's manual. The surgeon-scientist John Freke (1740) thought that electrostatic charges were involved. But an a priori sensitivity, or 'irritability' as it was called, was the most popular explanation. When the sequence of events which leads to the sudden infolding of the leaves was finally mapped in the late twentieth century, it involved fluid movement controlled by electrochemical signals – but not in the simplistic ways envisaged in the eighteenth century.

In 2013 the Australian plant scientist Monica Gagliano devised an experiment which revealed previously unknown depths in mimosa's sensitivities. She potted fifty-six similar plants, and rigged up a system to give them a standardised trauma. The pots were dropped to the ground from a height of six inches every five seconds. Each 'training session' involved sixty drops. In the early stages all the specimens, to varying degrees, shut their leaves in the usual manner of their species. But some began to reopen after only four or five drops. By the end of the session almost all were ignoring the drop stimulus. Were the plants suffering from a kind of vegetal fatigue? Apparently not. When the same individuals were shaken manually their leaves immediately closed. But put through the drop test again, none of them closed. Gagliano repeated the experiment, using the same 'trained' plants, after one week and then after a month. Again they ignored the drop trauma, suggesting they had 'remembered' what they had learned, that the drops were no threat to them. Bees, in analogous experiments, forget what they have learned in forty-eight hours.

Whether this is evidence of a kind of vegetal intelligence is a matter of semantics. If humans were put through stress tests and discriminated between those that were dangerous and those that were not, remembered the difference and learned to act accordingly, we'd have no hesitation labelling the behaviour intelligent. But we're stuck with the idea, the prejudice maybe, that intelligence is inextricably linked to self-awareness, to conscious reflection and reasoning, and baulk at the idea of ascribing it to unconscious beings. This is odd, given how many smart human skills – the ability to walk upright, for example – are exercised instinctively. A plant's lack of a brain and a linked nervous system complicates things further, leaving us without much clue about how plants communicate, do calculations, store experiences. But of course we don't have much idea of how brain function generates consciousness. The philosopher Daniel Dennett has coined the term 'cerebrocentrism' for our reluctance to have a more inclusive sense of mind. Monica Gagliano suggests that 'brain and neurones are a sophisticated solution but not a necessary requirement for learning'.

Scientists working in the field suggest that plants may have a kind of 'distributed intelligence', that their millions of connected sense receptors act like a swarm, a group mind where the whole is more than the sum of the parts. An individual plant may be a closer analogy to a termite colony than it is to an individual ant.

*

I find I'm most impressed by *opportunism* in the vegetal world, when plants cope with or even cash in on situations entirely unfamiliar to them. In 2015, what is probably the most venerable tree in Europe performed a remarkable leap for survival and transitioned, became 'sexually ambiguous'. The Great Yew at Fortingall in Perthshire is a male tree, yew being a species where male and female flowers normally appear on separate trees. For what may have been as long as five thousand years it has produced pollen-bearing male cones, and generated no descendants in the immediate vicinity. In 2016 three female flowers appeared on one of the upper branches, followed by three bright red berries, presumably as a result of self-fertilisation. Conifers do this on rare occasions, but it's instructive to contemplate what may have precipitated this emergency measure at Fortingall. The Great Yew became a celebrity in the eighteenth century, when it had a girth of more than fifty feet. It was appropriated as a Druidic relic by antiquarians, and became the target of souvenir hunters, who hacked pieces from the already hollow trunk until it had effectively turned into two separate trees. By the nineteenth century the gap between them was wide enough to carry a coffin through. Memories of this attrition may be the reason why the tree – now further collapsed and shrunk to about half its earlier dimensions – is surrounded by an iron cage. But the effect of this has been more to keep the tree in than possible assailants out. Every branch had been lopped back to the fence line. In the wild, the yew's long, drooping branches root where they touch the ground, providing another means of reproduction beyond seeding. Cutting them back on the male tree at Fortingall effectively emasculated it. No wonder it eventually dug deep into its ancient epigenetic resources and came up with such a transgressive solution.

The cleverness of plants is remarkable in its own right and ought, I feel, to inspire more respect than regarding them selfishly as mere 'service providers'. The challenges they grapple with – multi-sensorial communication, novel means of self-propagation, negotiation with predators – are common to the whole biosphere, and we would be wise to learn from them.

On one of my Chiltern wanders, I found myself deep in a beechwood and confronted by a huge picnic table, high enough to shelter under, and painted with a blue-and-white-check tablecloth. It was like a prop from Alice in Wonderland. *A couple of hundred yards further on was the skeleton of a small tree, welded out of already rusty iron and carrying, in place of leaves, individually cast metal fish. I sat under it for a while, and had a glimpse through a mesh of green leaves and iron carp of a hang-glider passing against a blue sky, an extraordinary blending of elements. It turned out that I was in the Chiltern Sculpture Park, one of the Forestry Commission's experiments in mounting trails of artworks inside the environments the sculptures were exploring. For the next couple of hours I wandered about not always sure whether I was seeing an art object or a moment of woodland serendipity. Dead trees were propped up in the crucks of the living. Was this pure chance or a sculptural metaphor for the interdependence of all components of the forest? Were the regular white gashes on beech trunks some runic inscription or just the result of squirrel nibblings? The most evocative, obvious artwork was a long, gently curving rib of wood supported about three metres above ground level on wooden trestles. Its thin, tentative – I'm tempted to say pencilled – horizontal line seemed simultaneously to emphasise how much woods are places of strong vertical growth and to conjure up the horizontal pathways of woodland creatures. The best woodland sculpture can change and refocus your attention to the living processes in a wood, and at the same time comment on the relations between human and natural creativity.*

I was first invited to write about landscape and art by the journal Modern Painters *(edited by Peter Fuller) in the early 1990s and over the years got to know, review and sometimes write catalogue essays for many artists, including Andy Goldsworthy, David Nash, Kurt Jackson and Maurice Cockrill.*

Andy Goldsworthy

First published in the Guardian, *2007*

It passed through the Yorkshire Sculpture Park's restaurant as smoothly as a held-aloft cheeseboard, its ripe flower buds winking against the dark wood. Andy Goldsworthy had ordered up another blackthorn branch, and it was being ferried from a bush just outside the restaurant to the gallery where he's working. If you followed its progress you would pass scenes almost medieval in their industry, but brushed by post-apocalypse fiction: rooms full of artisans and apprentices, some dressed in biohazard suits, snipping hair, moulding clay; rooms like Stone Age burial chambers, their slate domes and gravity-defying oakstacks echoing in the bare space. And then the gallery containing the sculptor himself. Goldsworthy is standing on top of a cherry-picker, pinning stalks of horse-chestnut leaves together with thorns. Stretching right across the room and up to the ceiling is the lacy curtain he's weaving with such wilful primitivism, his spiderman's web. It's almost finished, and is already warping the gallery's sense of space. You're in a man-made interior, but you could be in a tree, gazing out through this two-dimensional canopy. You have either the chestnut screen in focus or the blank wall beyond.

I watch Andy as he presses the start button and glides slowly down to earth, and think what a perfect image this is of the ambivalence that's at the heart of his work: the Neolithic craftsman riding his hi-tech mount; the delicacy and perishability of the swaying screen against the monumentalism and sense of history of the stonework. It's a while since I last talked with him, but this isn't the moment for heavyweight probing. He's more concerned about his hands, and holds them up towards me like a prizefighter. They're smothered with plasters and nasty-looking sores from the vicious, inch-long thorns. 'Is there *arsenic* in them?' he asks, only half jokingly, hoping I might have some botanical eureka to explain his suffering. I reassure him that this is what blackthorn always does, as generations of hedgers have found. *Of course* Andy made his pins from blackthorn, the slaathorn of his native Yorkshire, the bush of the bitter sloe. They do the job and remind you of the pain of labour at the same time.

The other thing he's worried about is whether the show will be finished on time. There's another whole installation to be built, two sheepfolds to be rounded off. This exhibition, to mark the thirtieth birthday of the Yorkshire Sculpture Park on the Bretton estate near Wakefield, is the biggest he's put on in Britain. I've counted more than sixty people working out in the mud and March drizzle: student volunteers from Bradford Art College, where Andy himself studied in the 1970s, his two full-time master masons, local drystone wallers, estate foresters. I doubt there's been such communal creative fury in West Yorkshire since they put up the local cathedral. The exhibition also marks the twentieth anniversary of Andy's seminal residency in the park in 1987, and it might be tempting to call it a retrospective – except that a retrospective is the one kind of show that can never fully be put on by an artist in thrall to fleetingness, to the exactness of place and season, and to the sheer mutability of nature.

Perhaps a revisiting would be a better description. But for this you need the photograph, an uncomfortably stolid medium to fix the artist's epiphanies.

Goldsworthy's works are often themselves a kind of snapshot, a serendipitous glimpsing and grasping of fugitive moments of contrast and natural metaphor. It's a moot point whether the act of making or the photographic record of it is the true 'work'. Either way the photos of that first Yorkshire residency are here as part of the show, to remind you of how it all began. They demonstrate all Goldsworthy's basic raw materials and the motifs that were to become the signatures of his work: cracks in rocks arranged into a snake, autumn rowan leaves making a flaming iris around a black pupil, bridges – archetypes of solidity – constructed of ice, snowballs kept refrigerated until midsummer and then disgorging their seasonal cargoes of fir cones and spring flowers. Natural materials celebrated for what they are, but also for what they are capable of becoming, for what, in these transformations and reversals, they can be made to reveal about the commonality and differences of the world's stuff. One of his 1980s Yorkshire pieces was an angular grass line, wandering around the voluptuous bole of an ancient beech tree. He captioned it: 'Continuous grass stalk lines / each stalk pushed into the wider hollow end of another / or two thin ends joined with a short length of thicker stalk / edging a hole, climbing a tree / pinned with thorns.' If this sounds more like an instruction manual than a title, it's because the idea of *making* is central to his philosophy. The line relies on qualities that are special to grass, but could never occur just so in nature. And its form – like an elongated hieroglyph – seems to echo the human lettering of the graffiti on the background beech.

It's the extent of his own physical intervention that marks Goldsworthy out from other artists who see their work as 'a collaboration with nature'. The group which emerged in the 1970s under the loose title of 'land artists' share a belief in the sanctity of nature, and that an art which 'trod lightly on the earth' must reflect the transience and unpredictability of the natural world. Artists such as Richard Long, who exhibited written notes about walks he'd made, and Gary Fabian Miller, making photoprints of leaves hung in sunlight, were minimalists, cerebral hunter-gatherers, sometimes to the point of being purely conceptual. By contrast Goldsworthy is a canny dirt farmer, scrabbling, chopping, mud-slinging, sowing (and sewing), wall-building, wallwrecking – all a similar distance from 'treading lightly' as a handshake is from a stroke with a fingertip. Ironically, this traditional approach to the artist's role received somewhat muted respect from the art establishment at first. He was looked on by some as a mere craftsman, an exterior decorator, a tricksy designer, or – heavens! – someone who was just *playing*. This was the dismissal that hurt him most at first, but more experienced now and with children of his own, he is happy to acknowledge that aspect of his work. It is not the remotest bit disrespectful to say that the prototype of all his work is the daisy chain, the solitary 'day's-eye' made into a community, the soft plant tissue mimicking the forms of iron. Children were peeling leaves and welding ice millennia ago.

Goldsworthy's extension of these compulsive human habits is his way of exploring the essence of natural forms. His dandelion flower-chain is floated on twigs inches above shoals of bluebells (1 May 1987) – a plain-man's garland above the elegant ultramarine, a conceit that in reality only lasted as long as the flowers themselves. 'Movement, change, light, growth and decay,' he wrote ten years ago, 'are the lifeblood of nature, the energies I try to tap through my work. I need the shock of touch, the resistance of place, materials and weather, the earth as my source. I want to get under the surface. When I work with a leaf, rock, stick, it is not just the material itself, it is an opening into the processes of life within and around it.'

Which brings us to the black holes. These vanishing points, voids, still centres, occur repeatedly in his work, like a kind of colophon. There are black holes in cairns, in leaves, in hollow snowballs, in wheels of brushwood. 'A black hole will probably be the last work I make,' he declares with gallows humour. They earned his work some cynical jibes in the early days from feminist critics who thought the holes were symbolic of a dehumanised and feared vagina. Female and mysterious they may well be, but they are far from fearful, or alien. They ground his work. They are portals between the human and natural worlds. He tells me of the profound moment they became important to him. He'd been exploring some dunes, burrowing into them, teasing the wet sand. The sand dried to form a crust, and when Andy cut a hole through, the shell was so thin there were no shadows, just a tunnel of black between him and the untouched English desert beneath. 'For years,' he says, 'I hovered around the hole's edge, trying to find a way in.' Then, in Dorset in 1992, he made an installation called *Hard Earth*, plastering the inside of a room with white clay, so that it looked completely empty, dimensionless. But as the clay dried, it began to crack with black lines, and the room became not lined with clay, but made of clay. The clay's essential identity had invaded the human's space. The black had been 'released from its container'.

There's a clay room in the current show, the plaster reinforced – like old-fashioned wattle and daub – with hair. Human hair in this case. Andy's team trawled the hairdressers in Wakefield and Leeds over the Christmas rush to collect it. This is what the art students have been working with, cutting the tresses into inch lengths, kneading them into the clay, and then slapping it on the walls. Andy relishes the thought of 'all that DNA' – from Bradford teenagers' fingertips and West Riding blue-rinses – locked up in the walls, and then released from its containers as the cracks start to form.

If the inner energy of nature is one of Goldsworthy's basic preoccupations, time is another. Time passing, frozen, submitted to. Ephemeral moment and deep history. At the edge of the Sculpture Park, at the furthest possible remove from the indoor galleries, there's a brand-new work, *Hidden Trees*. You can't see it at first, and that's the point. It's been built in one of the Bretton estate's old ha-has, a kind of boundary trench popular with eighteenth-century landowners which, by making a fence unnecessary, provided the optical illusion that landscaped park and worked farmland were part of a seamless whole, and park and working field, squire and labourer part of a harmonious partnership. *Hidden Trees* challenges this complacency. A dry stone wall runs along the foot of the ha-ha, and in three places is broken by cavernous chambers. Great oak logs – the top branchings from trees felled on the Bretton estate – lie wedged horizontally across the space. They're stripped of bark and already as dark as fossils from the coal measures. Their trunks and branch tips are locked into – bound by – the drystone walls of the chambers. Looking down at them makes your head spin. The ha -ha was traditionally meant to disguise the differences between the cultural and the natural. Peer down into this one and you see exactly what was thought better hidden: the forest cleared for and handcuffed by the walled field. No simple propaganda here, but a severe reminder of the deep layers – natural and human – ingrained in a landscape.

The relation of trees and time has been a favourite Goldsworthy theme. In the south Australian bush in 1991, with no planning at all, he lighted on a dead mulga tree, decided to climb it and cover it with damp red sand. The sand dried and stuck, petrifying the tree a million years before history would have. His photograph of it is like a flame spouting from the desert. In Paris five years later he collaborated on a ballet, *Végétal*, with the choreographer Régine Chopinot. He wrote a visual storyline around the themes of tree growth, time and work, using the dancers' bodies as mobile sculptures. Thirteen of them, dressed for action in boiler suits, plus an assortment of stones, branches and leaves, created five narratives based on the growth stages of a plant. In 'Root' they shoot about the stage on their bottoms, almost colliding with each other but always changing direction at the last moment, until every possible combination of dancers has been used and every square foot of stage tendrilled over. In 'Branch' they weave a kind of enclosure from branches, dancing them onto the stage one at a time. Then they dismantle it, build a wattle pen, dismantle that, and finally make an even wider stockade in which the rest of the work is performed. The repetitions were compelling in what they said about natural recycling and the rhythms of growth. But not everyone agreed. I was there on the first night, and there were catcalls from some of the Parisian audience not prepared to see theatrical time give way to natural time. They were quietened during 'Leaf', in which the dancers become transpirers themselves, making their own rhythmic soundtrack by hyperventilation and intense ribcage workouts. Even in these dramatic moments, Chopinot's choreography gave bodily expression to the small, vernacular details of movement that are such an essential part of Goldsworthy's vocabulary: the curve of a leaf's fall, the way a child plays with a stick while carrying it.

An acceptance of natural time as a collaborator in the works means a willingness to be both patient and flexible. To wait for the temperature to fall low enough for ice to stick to rock. For the moment of perfect calm that enables a framework of knotweed stalks pushed to the bottom of Derwent Water to merge with its reflection into the form of a perfect circle (20 February 1988). In the current Yorkshire show the most endearing piece – 'Sheep Paintings' – involved waiting for flocks of sheep to become collaborators in the work. Goldsworthy had laid out large horizontal canvases in some pastures, with a sheep-lick at the centre. Over hours and days, the footfalls and dung droppings of the sheep built up a physical record of their activity on the blank canvas, a brown busy-ness, like a dense, modernist musical score. In the centre was this pure white void – a *white* hole for once – where the lick had stood: the sheep version of the village pump. They're engaging pictures, full of idiosyncrasies: a single trail of hoofprints, from a latecomer or a lost appetite, winds away from the throng on one side; on the other is the dark bustle of a feeding frenzy, the cheering of a sheep crowd.

But these pictures aren't exactly the copyrighted creations of the sheep. The management has had a say. Goldsworthy could well have shovelled up a slab of turf from under them and put that in the gallery, complete with their droppings, but he's not an *objet trouvé* artist. Everything he does has, at some point, the imprint of the human. 'My art is unmistakably the work of a person,' he insists, 'I would not want it otherwise – it celebrates my human nature and a need to be physically and spiritually bound to the earth.'

His other work with sheep in mind emphasises this. In one of the park's woods, a sheepfold or enclosure is laboriously rising. The beautifully grained sandstone, dug from local quarries, is being laid into a wall which will ultimately be more than two metres tall. It will have no entrances or portholes. Those of us who saw it being built know that in May it will be full of happenstance bluebells, but for most visitors it will be as blank and secret as a black hole. But this work, set bang in the path of a right of way, is not just about the mysterious innerness of nature, but about its misappropriation by humans, about the dual meanings of 'enclosure'. The wall can protect nature, or ride roughshod over it. It cossets one community at the expense of another. In *Wall That Went for a Walk* (Grizedale Forest, 1990), he subverts these notions, building a wall that snakes through the trees, following the lie of the land and incorporating the odd tree and rock into its length, rather than flattening them. In his 'Give-and-take' sheepfolds (echoed in installations in the new show) he creates the sculptural equivalent of symbiosis as an alternative to ex-closure. The first came about when he leased a small parcel of woodland in Dumfriesshire. Part of the agreement required him to build a dividing wall with the neighbouring farm. He included two sheepfolds in the wall: one opened on the farmer's side for his sheep, the other on Goldsworthy's side for his sculpture. The animals graze on the sculptor's land, and art flourishes in the farmer's. He has explored this device at a more overtly political level in an installation created specially for the National Gallery in Washington DC, just a few hundred yards from the White House. It's based around one of the glass walls of the gallery, the boundary between 'inside' and the world beyond. Outside are a group of Goldsworthy's familiar cairns, low-slung mounds like stone molehills. Except that two of them have broken through the glass wall, dissolved the boundary, become both insiders and outsiders, The day I saw it my car was docked for being parked too close to the seat of power.

What intrigues me most about Goldsworthy's work are the seeming paradoxes in it. His reverence for nature and his insistence on physical intervention. A passion for delicacy and ephemeralness that can relish the few minutes of existence of an ice pinnacle, coexisting with a respect for human works that have endured for centuries. Perhaps they are only paradoxes if you think of artists and viewers as outside nature rather than part of it. But when Goldsworthy says his work is unmistakably that of a person, what does he mean? It is always unmistakably the work of Andy Goldsworthy, of course, but is he suggesting where his work fits in the tradition of sculpture too? His sense of belonging with the natural world marks him down among the Romantics, though he has none of their intense devotion to human emotions. His fascination with ornamentation, with taking a natural form and encouraging it to proliferate, play, become exuberant, was one trademark of the carvers of Gothic art. You might even see him as a kind of abstract expressionist – except that what he is expressing is not entirely 'his' or even 'ours'. When I first wrote about Andy's work, in 1991, it occurred to me that many of his pieces had the look of *evolved* forms, and that the artist's relationship with his material was like that of a rogue twist of DNA or a benign virus with its host cell. He worms his way into a bundle of branches or a single leaf and sparks off new developments and mutations, new possibilities. Simon Schama suggested in his television series and book *Power of Art* that great art is about the exploration of what it means to be human. Goldsworthy has a more inclusive take on this in that he is exploring what it means simply to be alive, a thing in the world. No art can avoid stressing the differences between human and natural creativity, but Goldsworthy's also explores the similarities: how the ice sculpture echoes and emphasises the gracious formation of ice; how improvisation with leaves reflects the playfulness of the evolution of the leaf itself. In this sense Goldsworthy is a truly ecological artist, a founder member of a new tradition.

We sit, chatting over such meaty matters. The chestnut screen is on schedule and he's enjoying a breather. But the show itself is suddenly less advanced than it was. True to nature, so to speak, he's had another brainwave. Another form has come onto the artist's evolutionary conveyor belt. He's been watching the evening cleaners at the Sculpture Park, and become enthralled by the debris they're picking up: the shoe-mud and flaked skin, the bits of blackthorn bark and dropped food and wisps of sweet wrappers. He imagines himself as a human dung beetle, collecting it up every evening, all that evidence of consumption and travel and wastefulness, rolled up into a solid, uncompromising, highly visible ball. The team has been alerted. Volunteers are needed to make a nightly ball for the whole duration of the show. At the end of the year, Goldsworthy himself will roll them all into one, the biggest and most embarrassing globe of garbage on the planet. Then – he doesn't know yet – perhaps he'll let it rot, disgorging its contents like the summer snowball and the cracking clay wall. Or maybe it will turn into compost and catch fire.

In a Brown Shade

First published in the New Statesman, *2016*

The year 2016, though somewhat overshadowed by the anniversary of another English craftsman, William Shakespeare, marks the tercentenary of the birth of the landscape designer Lancelot Brown, whose supposedly naturalistic confections around more than 130 country houses earned him the popular tag of 'Capability'. The conjunction with the Bard has pleased Brown's supporters, who believe he should be similarly regarded as one of the 'curators of English identity'. 'He stands behind our vision, and fantasy, of rural England,' reads the blurb on one biography. Fantasy for sure. It's hard to see what connection the classic Brownian prospect – artificial lake, vast lawns, mansion framed by trees – has with real rural England, let alone 'nature'.

But hyperbolic praise for the great earth mover is nothing new. In 2011 the *Financial Times* carried a series of articles that suggested Brown should be 'canonised'. The historian Norman Scarfe wrote, startlingly, that 'England's most original contribution to the whole history of art lies in the landscape, and was an affair of creating harmonious pictures with the land itself.' Back in the eighteenth century Horace Walpole, a staunch defender, suggested that 'Such was the effect of his genius that when he was the happiest man he would be least remembered, so closely did he copy nature that his works will be mistaken.' This isn't damning by faint praise: Brown's champions have always claimed that his supposed ability to imitate nature was his special gift. But the plaudits open up the contradictions at the heart of Brownism. He's praised both as a uniquely original artist and a faithful copyist; as a minimalist and a land engineer on the epic scale; even as a rustic Marcel Duchamp, reimagining nature by dragging it into the gallery of the garden. In the end these aesthetic arguments are a matter of social solidarity and personal taste: anyone is entitled to make their garden into an installation. The more serious objection to Brown – or more properly to his cult – is that he is implicated in a powerful but false creation myth, of the kind the great historical ecologist Oliver Rackham dubbed 'pseudo-history'. This origin story posits Brown as Promethean, creating the style which, by a kind of trickle-down, remodelled the entire landscape of lowland England. The claims made on his behalf, when you step back, are of preposterous hubris. 'England's most original contribution to the whole history of art'? 'Creating harmonious pictures with the land itself'? Whose harmony? Whose land? The ethos of 'landskipping' has had corrosive effects on our understanding of what is natural, and of the millennia of landscape making that preceded Brown. It suggests that 'proper' rural landscapes have been, and should be, created by a planning process, on a drawing board.

The roots of Brown's designs lay not in nature but in the philosophy of 'Improvement' which pervaded public affairs in the eighteenth century. Improvement – in farming, industry, colonial adventure – lay at the heart of Whig philosophy and was an economic project cloaked in moral justifications. In landscape it involved presenting a clear view of a landlord's benevolence and entrepreneurial spirit together with a physical structure that dramatised the social hierarchy of the estate. The so-called 'English style' is supposed to have been a reaction against the contrived topiary and symmetrical parterres of French and Italian gardens, which represented the stiff intellectualism of the Continent, and was to be countered by buxom curve and round copse and Albion's indigenous 'emblems of pure legal liberty'. In truth both were domineering idioms, and the much-touted 'naturalism' in Brown's and others' designs has always been just spin. Rich men with access to the new technologies of surveying and drainage could now produce their own 'nature'. Alexander Pope's famous injunction in his *Epistle to Burlington* to consult 'the genius of the place' turns out to be the programme for an inquisition, where if the *genius loci* is found wanting the real genius, the landowner and his minions, must step in. Brown's apologists (he wrote little about his own work) agreed: 'the living landscape', Walpole instructed, must be 'chastened or polished'. Nature must be 'rescued and improved', the damage of the Fall repaired.

A central aim of English eighteenth-century landscaping was to demonstrate the standing of the Big House. A key device was the ha-ha, a sunken, walled ditch which excluded grazing animals while allowing the working countryside beyond to be seen unobstructed – 'taken in' as part of the view. These agricultural acres were themselves being transformed by the forces of Improvement, so that the landowner had, in the same field of view, both his aesthetic playground and his commercial workplace, though reminders of the less pretty realities of farming, like pigsties and workers' hovels, were screened by strategically placed tree-belts. On the other side of the ditch, ordinary folk, excluded along with the cows and deer, could see the seat of rural power in plain and unambiguous sight, often on a slight rise in the ground. Humphry Repton, Brown's successor, spoke approvingly of 'that charm that belongs to ownership, the exclusive right of enjoyment, with the power of refusing that others should share our pleasure'. Inside this overall structure were a series of pick-and-mix motifs – perfect parabolas of gravel, acres of lawn, water features where water had no business to be, follies inspired by neoclassical painting – all assembled without reference to site or historical context. Cumulatively they built up an entire grammar of exclusion and control.

Yet Brown himself was no high-born elitist. He grew up in Northumberland and served six years of apprenticeship in the gardens of Sir William Loraine at Kirkharle. Although he did develop a personal style, he's best understood as a factor, a man adept at giving physical expression to his patrons' ambitions. He worked on several commissions at once, sometimes simply modifying earlier layouts by designers such as William Kent and Charles Bridgeman. But despite the legend of harmoniously absorbing the local countryside his projects were often brutally destructive. At Croome Court in Worcestershire, he drained a huge 'morass' in the flood plain of the Avon (today it would be a protected marshland) into an excavated lake via artificial rivers. At Moor Park in Hertfordshire he created clusters of obtrusive hillocks, topped by groves of trees – 'a dullish piece of Hertfordshire', one of the owner's family remarked, '[transformed] into a very fair imitation of Italy'. So much for consulting the genius of the place. The Grecian Valley at Stowe, an estate where Lord Cobham had previously cleared three whole villages that spoilt his view, was made by moving earth in the opposite direction – 24,000 cubic yards of it, dug out and carted by hand. Lord Milton had the entire town of Milton Abbas in Dorset razed to the ground while Brown was working for him between 1763 and 1770, replacing it with ten miles of straight carriage drives and plantations of pines.

Tree planting was a significant ingredient of Brown's improvements. Trees were territorial markers, visible demonstrations of status, emblems of continuity. They could conjure up a mythic pleasure ground or a valuable investment. Different species had different symbolic values. Oaks and elms were icons of patriotism. Imported evergreens like cedars of Lebanon and Douglas firs were favoured because they added an aura of Britain's imperial power and boundless capacity to import new wonders from across the globe. Saplings were used, but many patrons wanted mature trees, to achieve 'the Immediate Effect of Wood'. Brown designed a 'planting engine' for this purpose, a flatbed mounted on two large iron wheels. I have an engraving of a six-man machine carrying a huge beech tree. It looks like the triumphant parading of a captured gun. Brown's favourite tree motif was 'the Clump', a woodland canapé, suggestive of nature's mysteries when viewed from the terrace, but purged of any real suggestion of wildness such as the undergrowth which forms the natural transition between wood and grass. Uvedale Price, Brown's most articulate contemporary critic, wrote: 'the *clump* – a name, which if the first letter were taken away would most accurately describe its form and effect ... [is] like so many puddings turned out of one common mould.'

Planting amenity trees seems so self-evidently a force for good that it's hard for us to understand what a novel practice it was before the eighteenth century. Why bother when trees appeared so magnanimously of their own accord? Now, addicted to the practice as a ritual of atonement, we've become blinkered to the fact that it's yet another expression of human power over nature. That trees have perfectly adequate, fine-tuned reproduction systems of their own seems to have vanished from popular understanding.

Brown's gardens can of course be captivating, in a hypnotic way. A few years ago at a literary festival at Petworth House, Sussex (grounds laid out by Brown in 1753), I had the rare perk of dining in the great hall, looking out over continents of lawn that seemed to extend unbroken all the way to the Midsummer Eve's sunset, and with J.M.W. Turner's original watercolours of the burnished scene just behind my chair. It would have been ungracious and shallow to have been anything other than enchanted. But in the colder light of the following morning I realised I might as well have been gazing out over a suburban playing field. Half a lifetime and many Brown landscapes further back I lived close to the great wastes of Ashridge and Berkhamsted commons in the Chilterns, a dynamic, storytelling prospect of heath and ancient beechwood which had been evolving under common usage for more than a thousand years. All that spoilt it were the remnants of Brown's tinkerings in the 1760s – weird triangular glades cut into the woods and tightly mown rides forced through the natural maziness of gorse and beech for no other reason than to reveal the ghastly neo-Gothic prospect of Ashridge House. What they said to me – as they said to Brown's contemporaries like Price, William Cowper and Oliver Goldsmith – was that his designs have no meaning beyond their demonstration of power. They're decorative ornaments, devoid of ecological pattern or deep historical grain. There has always been another way large-scale landscapes have evolved. Classic prospects such as the Yorkshire Dales, the Wye Valley and the New Forest developed out of a ceaseless negotiation between natural forces and human ingenuity which stretches back at least five thousand years.

Alas, the heirs of Improvement haven't wised up to benign natural processes, nor developed any respect for millennia of vernacular landscaping. In 1979, in the wake of Dutch elm disease, a group of designers published an influential manifesto called *After the Elm*. Praising the example set by Brown and his ilk, and the 'informal, apparently almost casual ... English style' created in the eighteenth century, they set out a Soviet-style plan for a national 'Grand Design' in reparation: 'it is therefore more important than ever', they wrote self-promotingly, 'that the initial plan is prepared by professional Landscape designers, and is strong, cohesive and convincing, not a collation of haphazard suggestions by many people' – ignoring the fact that unplanned, communal enterprise is what gives the English countryside its variety and character.

The nearest large-scale Brown landscape to my home in East Anglia is at Heveningham Hall in Suffolk, which he began to lay out for Sir Gerald Vanneck in 1783. Driving to it from the west you pass through the upper Blyth Valley, an intimate pocket of twisting medieval lanes arched over by ancient hornbeams, Suffolk's county tree, and bordered by tiny water meadows. To move up the road half a mile and come upon the estate is like being teleported to a different continent. Beyond the park's iron railings, immense acres of grass and countless sheep sweep up to the hall in a grand statement of authority. Brown created a dramatic lake over a mile long and preserved a few pre-emparkment oaks, but they stand abruptly in shorn grass like municipal statues. Heveningham Hall was bought in 1994 by the property magnate Jon Hunt, who commissioned the landscaper Kim Wilkie to complete Brown's design. Wilkie simultaneously worked on what is called 'the Wilderness Reserve' on the estate, a well-meaning project but a long way from any normal idea of wilderness or the 're-wilding' it claims to be. Echoing Brown at Croome, he drowned a wet alderwood under a lake. He planted 800,000 trees in regimented lines and hard-edged plantations, devoid of wild undergrowth. Seventy-two thousand were ash, which Wilkie, fearing ash die-back disease, removed and replaced with oak and hornbeam – 'an £850,000 mistake', he judged. Indeed it was. Thirty-five miles to the south-west, next to the wild cherry-and-hazel wood known as Arger Fen, the Suffolk Wildlife Trust let a large arable field go feral ten years ago. It turned spontaneously into an ash wood. Then, when many of the ash seedlings fell victim to dieback, into a self-sprung wood with ten different species of tree distributed entirely naturally across the site – all with no cost or planting contortions whatsoever.

The Serendipitous Garden

First published in Gardens Illustrated, *2016*

Some years back, whenever we had mussels for supper, we were in the habit of hoicking the remains onto the compost heap, thus injecting, we reasoned, a useful shot of calcium into the nutrient chain. But the garden's *genius loci* had other ideas. In the days that followed each dumping, small clusters of shells began to appear all over the place. It wasn't some random strewing. They were arranged in neat cairns at the foot of trees and tucked against shed walls like bowerbird displays. I've no idea who the landscaper was. Magpies make ornamental collections, but not usually on this scale, and I'd plump for a particularly aesthetic brown rat. Whatever lay behind the ornamentation, the lustrous blue and mother-of-pearl waymarkers were something we would never have dreamed up for ourselves.

I like the idea of the garden as an arena for other organisms' creativity. We tend to assume that gardens are, by definition, expressions of the gardener's tastes and ideas. But why can't they also be open stages, frameworks inside which wild cards – uninvited organisms, dramatic weather, ancient processes of succession and decay – improvise their own landscapes?

This is how a good deal of our plot in south Norfolk operates. There's a touch of serendipity even in its structure. The two acres stretch from a formal herbaceous border at one end to a small wood at the other, and degrees of intervention follow the same gradient. At the juncture of cultivation and the wild there is a deep pond, fed by a cold spring. I'm not sure it was deliberately created as a body of water; more likely it was dug four hundred years ago to provide clay for the house walls, and then filled up of its own accord. Now moorhens nest in the reeds and tits and woodpeckers drink under cover of the tangled tree roots that drape its steep sides.

Trees enclose and permeate the garden and are what everything in it would turn into, given the chance. A good few came with the property, and our most gleeful house-warming present to ourselves was felling nine monstrous leylandii in the first week of occupation. The fifteen mature hardwoods at the far end of the garden are another matter. They are about fifty years old, well spaced and mostly of native species – cherry, ash, beech – and I could see the beginnings of a wood there. With hindsight I'm not sure why I bothered to plant a few saplings in the gaps, when it was quite obvious that the woody incumbents would do their own propagation. Sure enough, freed from the mowing that our predecessors had done between the trees, the ground soon bristled with self-sown seedlings. Oaks came from the big tree on our boundary, hybridised here and there with pollen from turkey oaks in the lane. Cherries sprang from suckers and stones deposited by blackbirds. Half a dozen other species, including hazel, hornbeam and rowan, had their fruit blown or ferried in from nearby hedges. A sweet chestnut shot up through a bramble patch. Ten years on we have a closed canopy and a two-layered understorey.

It's the ashes that are the worry. Sadly, they're the species which seeds least successfully in our garden, which means we haven't got the diversity of trees which might guarantee survival when ash dieback hits. We're only ten miles from the wood (called, ironically, Ashwellthorpe) where chalara first appeared in the wild in 2012, so we'll be lucky to escape. We have eight mature specimens, so if a significant percentage is afflicted it will change the whole feel of our home patch. The dominant tree, the super ash, is a multi-stemmed specimen rooted at the edge of the pond, and has a canopy twenty metres across. It's an amphitheatre for birds, and a dramatic weathervane. Ash branches are fluent, and when a good wind blows it is as if waves of wood are breaking over the water. There is nothing to be done about dieback, so my response is to take care of the self-sown hazels and oaklings that are growing close to the ashes, by lopping some of the latter's branches where necessary. That way there will at least be some established trees to grow into the skeletons of the dead ashes.

They can't replace the ashes; but then again, no wood is stable and each is continuously changing its species quota. In ten years, if we did absolutely nothing, the entire garden would become an unpredictably mixed woodland – which is a pleasant prospect for when we're too old to push the mower. The shrub layer and the ground vegetation will be a sight to see too: clumps of cistus and sage, relics of my Mediterranean garden – at least before the canopy closes over them; shade-tolerant meadow plants such as knapweed red campion mixed with brisk and opportunist annuals escaped from the borders wherever there are patches of open ground. It would be a garden able to spontaneously adapt to climate change, by using the plants of the diverse habitats and weathers that were established there.

The long-term inhabitants of a garden don't just reflect current horticul-tural tastes, but all that the gardeners – and nature – have done there in the past. When I lived in the Chilterns, a mysterious rambler appeared among the old roses one spring. It had elegant pinnate leaves and very few spines, and months passed before I nailed its identity as cut-leaved bramble, *Rubus laciniatus*. This is a teasing species for botanists of a historical inclination, because no one has the remotest idea about where or how it originated. I wondered how it had reached my garden until I remembered where I'd seen it before, growing on a heath behind a Suffolk cottage I'd lived in a few years previously. I'd made jam from the fruits, brought a jar back to the Chilterns, thrown the fermenting contents onto the compost heap, and the compost onto the roses. The bramble seeds had survived cooking and translocation and rot, and the comely bush that ensued became a small memorial to my own migrations and habits.

Our Norfolk garden has more than 150 species which were not deliberately planted, a good few with comparable origin stories. After Polly and I returned from a holiday in Crete, an exquisite dwarf campanula, endemic to the island, shot up in the herb bed. A couple of years later a nest of blue-flowered snakes uncoiled on the gravel, eventually revealing itself to be the Mediterranean annual, small-flowered viper's-bugloss, *Echium parviflorum*, whose seed I guess we must have picked up on our boots hiking in Provence the previous summer, or brought home in a box of figs. As for double-flowered greater celandine, it seems to follow me wherever I go, after I took home a seed pod from Kew Gardens. Increasingly I find that these incomers (I refuse to call them weeds) are far from happenstance; that we, as users and workers of the garden, are generating them – *cultivating* them if you like – by our behaviour and personal affections. They reflect the people we are, our histories and hoardings, the walks and holidays we take.

Some have deeper historical roots. Tansy, feverfew, evergreen alkanet (which crops up all over the garden – in the potato patch, the paths, and studding the white lace of cow parsley with cobalt buttons) were here when we arrived, and may be relics of old herb garden cultivation. The earliest map of the property I've been able to trace is from the early nineteenth century and shows a smallholding with rows of fruit trees at the front and a pond at the back. What is now our meadow was the corner of a large field labelled 'Hempland': the two bachelors who lived here at the time were farming cannabis – except that this was a non-psychoactive variety used for making linen. The cloth from this stretch of the Waveney Valley was renowned and sent to Kensington Palace and the Russian Embassy. I thought it would be a nice tribute to grow a patch of the fibrous variety on the land it had occupied two centuries before, so applied to the Home Office for a licence to raise this proscribed herb. Their frosty reply made it clear that home-growing was precisely what legislation was intended to prevent.

In the end the lack of a licence proved irrelevant. One warm summer a single cannabis plant appeared of its own accord in the herbaceous border, its famously five-fingered leaves waving mischievously between the phlox and the clarkias. It grew to about three feet high, put out its dull yellow flowers in autumn, and keeled over with the first frost. I'd like to think it was a long-dormant descendant of the crop our nineteenth-century small-holders grew for the London linen trade, but most likely it had sprung from bird seed, like the millet nearby.

Many of the incomers, ancient and new, put on more vivid displays. Inside the walled garden, the beds are subject to just the kind of regular disturbance gatecrashers relish. Headlands and knots of self-sown poppies, feverfew, marigold and datura edge the neat rectangles of vegetables, giving them the air of cubist paintings. They hang out there like popinjays, harmless, useless, but brilliantly and provocatively decorative.

The most dramatic of our stowaways arrived via a Trojan Horse manoeuvre for which we were entirely responsible. We have a house custom at Christmas of using somewhat eccentric growths as our seasonal trees (a wilding apple branch complete with yellow apples was 2016's). A few years back we noticed the wonderfully statuesque skeleton of a giant hogweed in the corner of a neighbouring field, desiccated enough to be no danger as a source of irritation (the green stems can raise blisters on the skin in strong sunlight). We carted it home like a captured flag and hung it with baubles. It looked strange and spectacular, a Yuletide triffid. On Twelfth Night we took it to the bonfire and thought no more about it. The following summer the unmistakable basal leaves of a hogweed, vast and jagged, appeared in the gravel directly outside the front door. Trimmed of its venomous lower leaves for the sake of callers it eventually grew so tall that its white umbels, the size of cartwheels, bloomed dramatically just outside our bedroom window. A couple of years later the whole cycle was repeated by the *back* door.

Giant hogweed was introduced to Britain from the Caucasus in the early nineteenth century. It was actively promoted by nurserymen as a cheap and hardy European alternative to the exotic wonders grown in the glasshouses of the rich. Packets of seed were sold for a few pence, and the hog became a hugely popular marvel. But like other vigorous species fashionable at the time, it escaped, and over the next hundred years spread across the UK, especially in damp waste patches. It is still very local, but has been declared a statutory weed, illegal to plant out deliberately in the wild. Some conservationists vilify it as an 'invasive alien', and I have some sympathy with them when it invades fens (very rarely) or native riverside vegetation. But arrived under its own steam in a garden it makes a dramatic statement about the autonomy, the energy of plant life. Vivid, luxuriant, growing twelve feet tall in a few months, it is the biggest and most architecturally striking herbaceous plant growing wild in Britain, and a reminder, in its current status, of how fickle our aesthetic tastes are.

Many life forms that penetrate the complacency of the ordered garden can upset the hierarchies of value we impose on its inhabitants. Most early autumns a curious yellow curd appears scattered across our meadow. The patches are frothy and shapeless and look more like some excremental fallout than anything living. It's called 'dog-vomit' fungus, but that's an insulting misnomer. I prefer to be reminded of lightly cooked scrambled egg. And it's not a fungus or even any kind of plant, but a slime mould, a kind of organism that is now on a branch of its own on the great Tree of Life.

Slime moulds are astonishing creatures. They spend much of the year as a swarm of microscopic single-celled animals browsing on yeasts and bacteria in the grass. Then in the autumn, or whenever there is an abundance of food, chemical messages sent through the air trigger the cells to come together and form the communal yellow splodges. At this point the aggregates meta-morphose into something resembling a fungus, putting up fruiting bodies and dispersing spores, so that the whole cycle begins again.

They have one other extraordinary property. During aggregation they form critically efficient channels between sources of food. In one recent experiment, scientists laid out oat flakes (a favourite slime snack) in the geographical disposition of the stations on the Tokyo subway system. The slime mould swarm scoped the whole area, then gradually began to refine the links between the flakes (along which food is shared) to echo the tracks of the underground train lines, sometimes finding even more direct links.

Slime moulds seem to me a metaphor for the social and ecological life of gardens. Plants and creatures and ideas swarm about in the territory and are then brought together in these foci of activity, come into flower and seed and ensure their own continuance.

In the autumn it is the turn of young human mammals to be the unpre-dictable garden intruders, the slime mould in its cell stage. Grandchildren swarm in the wood in this season, climbing trees, making fires and devising incomprehensible feats of orienteering and tricks with plants. Most of all they love feral cooking. I've taught them how to make what we called twists when I was a kid, shaving the bark off an ash or hazel twig, wrapping a ribbon of unleavened dough round it and baking it in the embers of the fire. Their patience and manual dexterity are not always great (their ages are six to twelve at the time of writing) and their confections are often gobbled badly burnt or barely cooked, and covered in a relish of dead leaves, soot and soil. They eat them with an indiscriminate gusto unrecognisable from their indoor eating habits. Now they're moving on to their own inventions, stuffing Bramleys with garden berries and golden syrup and baking them in silver foil.

We've not yet fire-seared the horse mushrooms and parasols that sprout under the cherry trees, but equivalents of the slime mould's fruiting stage are resplendent everywhere. Lichens effloresce on branches in the autumn rains, and the chrome-yellow *Xanthoria parietina* starbursts across the walls. Logs sprout bejewelled miniature toadstools – coral spots, dead man's fingers, clusters of minute bracket fungi whose identity still have me puzzled. As for the piles of wind-blown twiggery we stack up for kindling, anything can crawl out – wrens, hedgehogs, sexton beetles, sometimes a small child. Mostly, though, they just hunch down, an infinitely slow-moving statuary whose shrinkage shows that the capriciousness of decay echoes that of growth.

Autumn is the season in which our chief piece of actual garden statuary, the Hogtable, comes into its own. I'd long hoped for a bird feeder that was bespoke to our house, not something off a production line, and dreamed of making one from recycled bits of old farm machinery, so that we could enjoy the poetic irony of birds gorging themselves on relics of the agribusiness that had so damaged their populations. It was too much for my limited craft skills but we were blessed with a fortunate family connection here. The campfire-loving children's mum is artist Kate Munro, who was responsible for some of the wood and metal installations at the Eden Project. She came up with the brilliant notion of modelling the feeder on the giant hogweed that had graced our garden for the previous two years.

So for a week in summer she and her family became the artists in residence. She made the stalks from two debarked hunks of a sycamore that had come down in a gale. Then she sat among the meadow flowers welding together two giant umbels from an extraordinary collection of rural junk – spiral fence-posts, coil springs, harrow tines. When it was finished we installed it next to the pond, and it looked so handsome, so exactly right among the skeletons of cow parsley, that I could hardly bear to disfigure it with fat balls and peanut bags. The very first visitor, a few hours later, was a female sparrowhawk, who perched on an iron stalk, dreaming of prey to come ...

PART THREE

Common Ground

In 1967 the American ecologist Garrett Hardin published a controversial essay called 'The Tragedy of the Commons'. His thesis was simple but bleak. If a resource was used in common it would eventually be overexploited. He used as his paradigm the example of a notional grazier, who would always, by the supposed laws of self-interest economics, strive to put an extra head of cattle on the common. This action would be replicated by all other commoners, leading inexorably to overstocking and habitat degradation. Hardin extended his argument to cover all commons where limited space, resources, water, air or food was shared.

His thesis had a baleful effect at the time, and was used as an argument by conservatives not just against commons per se, but in support of the privatisation of public assets and communally owned resources. It has since become notorious for its abstraction and generalisation. His theory may have some relevance for so-far unowned resources like the Internet, or the atmosphere, but not for actual terrestrial commonland. Most commons, in this sense of shared tracts of land, have ancient and complex laws, written or oral, which regulate their use by the commoners and general public. Special local courts deal with breaches of the rules in many places. Where these don't exist mutual respect and reciprocity between commoners is far more evident than in Hardin's dystopian scenario. And Hardin ignored the fact that most real-world cases of environmental degradation occur as a result of the appropriation or enclosure of common resources by outside interests, not from greed by the commoners. Increasingly, the natural inhabitants of commons are being looked at as commoners too, with their own ancestral customs of usage.

From as long ago as I was allowed out of the garden on my own, my environmental politics has been shaped by the idea of commonality. My first outdoors adventures with tree-root dens and tribal rituals were played out in the grounds of an abandoned mansion, not literal commonland, but used by all the locals – children especially – as if it were. Only when it was fenced off for a new housing estate did the despoliation begin. North of the town were the great sweeps of beechwood and gorse heath of Berkhamsted Common, where an act of illegal enclosure by the Lord of the Manor in 1866, which would have likely led to the ploughing up of the site, was defeated by a historic piece of direct action, involving the throwing down of the fencing.

In 1977, I was invited by Nature Conservancy Council (NCC), then the government's advisory body, to write a book for a popular audience 'to widen the public debate about nature conservation in Britain'. I would be given full access to NCC staff and properties, but the book could be my personal interpretation. It was a huge opportunity and a privilege, and I spent two years touring Britain meeting scientists, NCC officers and land-owners, and visiting glorious habitats like North Meadow, the great field of fritillaries near Cirencester. It was like an extramural master's course in conservation and ecology. I called the finished book *The Common Ground (A Place for Nature in Britain's Future?)*, as its central thesis was that nature could neither ethically nor practically be regarded as a private resource. Only recently I learned that there were voices inside the NCC that didn't warm to my stance and attempted to have the book cancelled. They were apparently outvoted. The book was published in 1980 without any censorship, and became a set book on some university conservation courses.

I must have had more allies than enemies in the organisation as two years later I received a call from Michael Heseltine, then Secretary of State for the Environment, to join its governing council. The council was not a decision-making body. It was intended to be a link between government and NCC officers, to discuss priorities and principles, and generate the context (political and financial) inside which the workings of the scientific and policy staff operated. There was usually a smattering of big landowners on the council, and some academics and businessmen. I was appointed as a kind of representative of the lay public, and (my only specialism) to advise on communication tactics for an organisation anxious to connect more meaningfully with the public. Under the prevailing Conservative administration this was a difficult time for nature conservation, and I saw at first hand the ancient but still bitter debate between public and private interests. There were dark whispers that the government was considering selling off national nature reserves and fierce arguments about the ethics of compulsory purchase, which looked like the last chance for the increasing number of threatened sites, though it never happened. I also saw the beginning of what would become a critical issue for the NCC, the commercial afforestation of the Flow Country in northern Scotland, Britain's most important areas of blanket bog. NCC scientists were adamant that the march of the conifer plantations should be stopped, by legislation if necessary. There were a few victories. Some areas were saved, and the tax breaks that gave private investors an incentive to put their money into commercial forestry were closed. But the animosity towards 'English busybodies' among Scottish landowners and forestry interests and even, extraordinarily, some opposition politicians, grew until it became a campaign against the NCC itself. In 1991 the hostile forces won, and the NCC was emasculated by being broken up into smaller country-based units, and having its scientific staff hived off. I had left the council in 1986, and no longer having to keep up a position of impartiality, was able to accept an invitation from the *Sunday Times* the following summer to write a report on the whole Flow Country calamity. My first excursion into investigative journalism is published below as 'The Battle of the Bogs'. The pieces that precede this are profiles of friends and writers who helped shape my political views.

The Green Language

Originally a BBC Radio 3 Essay, part of a series celebrating the fiftieth anniversary of Silent Spring, *2012*

The publication of Rachel Carson's *Silent Spring* in 1962 was upstaged for me by the Cuban Missile Crisis. I was at Oxford studying philosophy and politics, and as chairperson of the university CND was heavily involved in the peace movement. It didn't seem the right time to be bothered about threats to the natural world. I'd been a birdwatcher and would-be nature writer in my teens, but those interests seemed a diversion beside the compelling need to oppose the threat of nuclear war. It wasn't that I'd begun to regard nature as of minor importance but that nuclear weapons threatened the annihilation of *everything*, and it would have been perverse to pull out one strand of the planet's life for special favour. At the height of the crisis I remember coming down from my desk in the Bodleian Library, standing on the pavement and gazing due east, wondering what it would feel like to see the mushroom cloud rising over London. My home lay exactly midway between Oxford and the capital, and I felt a fierce pang not just for my family but for all the insignificant objects I might never see again: the box hedge in our back garden, the single street lamp outside the house, all the lanes that lay between us.

Crisis does sharpen your sense of the extraordinariness of small moments. I still have an intense memory of what I was doing that final evening when we thought the world might end. I remember it – and *need* to remember it – as teatime on a Sunday, a moment of symbolic English domesticity, and have no wish to check the details of when it exactly occurred. Like many others I found I needed to do something mundane and normal to challenge the unreality of what was happening. I'd bought a pot of jam, a few planks of wood and some nails and taken them back to my one-room digs near Magdalen Bridge. And while I listened to the news, I made toast and began to put together a ramshackle bookshelf, a bastion against chaos.

Silent Spring was not one of the books that would be arranged on it the following, blessedly reprieved morning. I'd barely heard of it, and at the time wouldn't have properly understood its message that threats to nature were also threats to humanity, or that weapons of mass destruction and toxic pesticides sprang from a common well of human arrogance and corporate greed.

I discovered Carson's book – and refound my love of nature – a few years later. Even before I read it I was struck by the metaphorical power of that two-word title: the shared world of natural music silenced in the season of new life. It seemed to suggest so much – that there was a cultural problem here as well as a biological one, and that indifference to the complex and beautiful lives of our fellow creatures was a kind of barbarism. *Silent Spring* is one of those few books which can be said to have shifted the course of history, and even fifty years on it is hard to read its unflinching account of the chemical Armageddon that in some parts of the United States came close to wiping out all wildlife. It details the reckless way these poisons were spread about – sprayed on farms from planes, dumped in rivers, shovelled on lawns – with almost no consideration of their long-term and long-range effects. The crux of Carson's argument was that these were *persistent* poisons. They lingered on the grain, were concentrated in the livers of birds which fed on the seed, were accumulated in predators which ate the small birds. This makes the book an ecological treatise, about the interlinked chain of life. It details, quite devastatingly, how the poisons sprayed across America to kill the beetles which transmitted Dutch elm disease, for example, also killed the other insects which fed on elms and killed the birds which fed on the insects. The residue was carried on the wind or washed into rivers, where it eventually entered the bodies of fish, and thus found its way into our own bloodstreams.

I found Carson's insights and arguments prefigured in the work of another writer new to me then – the nineteenth-century poet John Clare. A century and half earlier Clare had also glimpsed the impoverishing effects of agricultural intensification. He was brought up in a poor rural family in Northamptonshire, and lived through the disorientating experience of seeing his home landscape enclosed by an Act of Parliament in the 1820s – an early step in a relentless process which culminated in the toxic chemistry of *Silent Spring*. The best of Clare's poetry, with its brilliant detailing of the intricacy and dignity of nature, is, like Carson's book, an attack on the abuse of the whole web of life.

The situation in 1960s Britain wasn't quite as bad as in the USA, but we still hadn't woken up to either Carson's or Clare's warnings. There were poisoned skylarks dying in the wheat. Peregrine falcons which ate small birds were on the fast track to national extinction. Vast new arable fields which had been created by the violent tearing-out of medieval hedgerows had their crops 'sustained' by up to a dozen chemical sprayings a year. Tracts of woodland that had existed since the end of the Ice Age were being obliterated to make way for short-term forestry plantations, sometimes by the use of defoliant chemicals. The countryside was being wrecked before our eyes (it still is), but it was too culturally shocking to admit that the destruction was being generated *inside* agriculture. Belief in the 'good stewardship' of those who managed the land was part of our national mythology, and it was far easier to blame some vague urban malignancy like 'development'.

I'd become a writer myself by then, and in 1977, when I was looking for an epigraph to introduce a book, *The Common Ground* (see introduction), it seemed natural to turn to John Clare, prophet of the commonwealth of nature. I chose a fragment of autobiography, about his childhood wanderings:

> I often pulled my hat over my eyes to watch the rising of the lark, or to see the hawk hang in the summer sky and the kite take its circles round the wood. I often lingered a minute on the woodland stile to hear the woodpigeons clapping their wings among the dark oaks. I hunted curious flowers in rapture and muttered thoughts in their praise. I loved the pasture with its rushes and thistles and sheep-tracks. I adored the wild, marshy fen with its solitary heronshaw sweeing along in its melancholy sky.

Clare had a limited readership in the 1970s. But five decades on he's become loved and respected as maybe the finest poet of nature in the English language, and an early warning voice about what we were doing to our planet. Even in the paragraph of prose above you can glimpse the way his language challenged the orthodoxies of both writing and landowning politics: the sense of locality in that dialect phrase 'heronshaw sweeing'; the curves and freedoms of the rising lark and circling hawk, so at odds with the fences and straight lines of the new rural planners; diversity – the rushes and thistles and curious flowers – thrown down as a challenge to monoculture.

Clare was no vague sentimentalist. He knew his natural history. He's officially credited with sixty-five first county records of birds and more than forty for plants purely from the compelling detail of his verse and prose. But it is in a kind of poetic ecology that his originality and importance lie, as a writer who could trace out the mutual dependence of things not just 'out in the world', but, as it were, in the very structure and syntax of his writings. Read his poem, 'Wood Pictures in Autumn', with its evocation of a landscape in which every element is interlinked:

> The woodland swamps with mosses varified
> And bullrush forests bowing by the side
> Of shagroot sallows that snug shelter make
> For the coy morehen in her rushy lake
> Into whose tide a little runnel weaves
> Such charms for silence through the choking leaves
> And whimpling melodies that but intrude
> As lullabys to ancient solitude
> The wood-grass plats which last year left behind
> Weaving their feathery lightness to the wind
> Look now as picturesque amid the scene
> As when the summer glossed their stems in green
> While hasty hare brunts through the creepy gap
> Seeks the soft beds and squats in safety's lap.

The phrases loop off the page like the unreeling of a fishing line. Again, there are dialect word and neologisms – shagroot, whimpling, plats, brunts – wonderfully onomatopoeic but also suggesting a world with its own rules, and vulnerable to invasion. Clare builds up a sense of the swampland's interdependence with a string of conjunctions: the willows *that* make a shelter for the moorhen, *into* whose lake a little streamlet runs, *through* choking waterweed, *while* – a key linking word in Clare's poetry – a hare pushes its way through a gap in the reeds *and* finds refuge there. The wet wood is not a static landscape, but a working ecosystem, connected by the movement and mutual usefulness of all its components.

And there's one other living organism here – the poet himself. Clare typically 'drops down' (a movement he also makes when jotting down phrases when they occur to him) and tries to envisage the scene from the point of view of the moorhen and hare. We can easily imagine him snuggling down in the reeds himself. So the bulrushes become a forest, and a willow's contorted roots protective enough to shelter a bird. Repeatedly through his work Clare expresses solidarity with other beings, and conjures an image of a kind of democracy of the fields – a contract which was broken by the authoritarian imposition of enclosure, and by all the aggressive agricultural specialisation which followed in its wake.

In a remarkable poem, 'The Lament of Swordy Well', he takes on the voice of an abused common:

> I'm Swordy Well a piece of land
> That's fell upon the town
> Who worked me till I couldn't stand
> And crush me now I'm down.

Swordy Well was a rough piece of grass heath on the edge of Clare's village of Helpston. Its fate was the familiar one of a multitude of so-called waste places. It was dug for sand and gravel, ploughed up for arable, turned into a more substantial limestone quarry, and in our own times used as a bike-scrambling track. Clare has the piece of land recount its own dismembering. The unstitching begins with the lowliest plants. The field flowers have gone, and 'The bees fly round in feeble rings / and find no blossom bye.' If a patch of clover peeks through in spring, it is ploughed up the next day. Then the bigger plants go, a process Clare explicitly links to ruthless profit-making:

> The silver springs grown naked dykes
> Scarce own a bunch of rushes

When grain got high the tasteless tykes
Grubbed up trees, banks and bushes.

And out with the rough grass and the bushes go the birds that nest in them, and the gypsies that camped among them. Clare is outlining here a chain of dependency between soil, plant, insect and mammal just as graphic as Carson's.

In this lament Clare has become unequivocally political, and many of his poems read like calls to arms against profit-driven assaults on nature. His vivid epithets of digging, stripping, grubbing, turning-over make the processes of agricultural intensification seem like acts of violation. In an elegy to an elm felled near his house, he addresses it in reverent old-fashioned diction, and coruscates the improvers who claim their rights trump those of all other living things: 'With axe at root he felled thee to the ground / & barked of freedom.' Barked of freedom! What an astonishing phrase, suggesting not only the bragging arguments of the land-owners, but what had happened to the tree itself. Clare writes that the elm owned 'a language by which hearts are stirred', and ends the poem (which was suppressed by his timorous publishers when it was first written) with an interrogation of the nature of liberty and a ferocious laying of blame upon those who silenced it:

> Such was thy ruin music making elm
> The rights of freedom was to injure thine
> As thou wert served so would they overwhelm
> In freedoms name the little that is mine
> & there are knaves that brawl for better laws
> & cant of tyranny in stronger powers
> Who glut their vile unsatiated maws
> & freedoms birthright from the weak devours.

In his battle hymn against enclosure, 'Remembrances', Clare likens the advance of the new agriculture in Britain to the violence across the Channel:

Inclosure like a Buonaparte let not a thing remain
It levelled every bush and tree and levelled every hill
And hung the moles for traitors – though the brook is running still
It runs a naked stream cold and chill.

The historian E. P. Thompson remarked that 'so close is the mutual ecological imbrication of the human and the natural [here] that each might stand for the other' and concludes that Clare 'may be described, without hindsight, as a poet of ecological protest: he was not writing about man here and nature there, but lamenting a threatened equilibrium in which both were involved'. In the end, the pain of the destruction of his home landscape aggravated a psychotic strain in Clare's personality, and he spent the last twenty-two years of his life in Northampton General Asylum. But even here he was not silenced, and continued writing.

The metaphor of Carson's title gives us a lens through which to view Clare's life and work. Clare was doubly alienated by his facility for language – from his fellow fieldworkers, and from the mute flowers and creatures that he loved. Yet at just that point where the act of writing seemingly distances him from the natural world, he joins it again, in the role of ecological minstrel. Of the many things his poems teach us – the value of the local, the mutual dependence of all living things – the best is perhaps that language may be our greatest ecological gift, and that the answer to the still-present threat of a silent spring is for us to sing against the storm.

Kenneth Allsop

Original version published as a foreword to David Wilkinson,
Keeping the Barbarians at Bay: The Last Years of
Kenneth Allsop, Green Pioneer, *2013*

At the time of his death in 1973, aged just fifty-three, Kenneth Allsop was a household name, a distinguished journalist both in the press and on television. Most weekday evenings he could be seen presenting the current affairs programme *Twenty-Four Hours*, an urbane, grey-templed figure touched with just a hint of world-weariness. He was a polymath of a kind, a jazz critic, novelist, dinner-party wit, and for many of us just coming into an awareness of environmental politics, a crusading Green elder and prose stylist. His first book, a semi-autobiographical novel (*Adventure Lit Their Star*) about the colonisation of outer London's gravel pits by a rare wading bird, the little ringed plover, had won the John Llewelyn Rhys Memorial Prize for 1949. His writing about conservation and the countryside was to become an increasingly important part of his life in the years that followed.

Kenneth was born in Yorkshire in 1920 and brought up as an only child by a mother who despised any open shows of emotion. He was, by all accounts, an introspective young man, a devotee of the nature writings of

Richard Jefferies and Henry Williamson (and of the latter's taste for open-top sports cars). He left school at seventeen and went directly into a job on a local newspaper. He was called up to the RAF in 1940. During the war he contracted surgical TB, had a leg amputated and spent two years in hospital. These losses – of a leg and a higher education – rankled into a sense of pain and resentment that stayed with him for the rest of his life, and may have been responsible for the intensity with which he lived and worked.

He was demobbed in 1946, and with his new wife, Betty, retreated to a bedsit in north London and a succession of hack journalistic jobs. His break came in 1948, when *Adventure Lit Their Star* was accepted for publication by Macdonald. It remains Kenneth's most heartfelt book, and it is hard to say whether he identified more with the indomitable birds which were scraping a living in the scruffy landscape of gravel pits and wasteland west of London Airport, or with the invalided RAF man who watched them. But the success of the book gave him the confidence (and money) to accept an invitation from friends in the Common Wealth Party to join them in an experiment in communal living in a manor house on the northern reaches of the Chilterns. He and Betty's two years at Barwythe were far from happy, or successful. Like so many communes, it struggled with internal disagreements about money, child-rearing, work-sharing and decision-making. Kenneth wrote his second book about the experiment, called *One and All*. It wasn't published until eighteen years after his death, and is revealing about the contradictions in Ken's personality, especially how a rigid and sometime authoritarian streak coexisted with an otherwise libertarian philosophy.

In the years that followed he built up a weighty journalistic reputation. He worked at the distinguished photojournal *Picture Post* for a while, and wrote book, theatre and jazz reviews for a wide range of papers. He also published several books, including a waspish appraisal of the cultural revolt of young British writers in the 1950s (*The Angry Decade*, 1958), and a history of the American hobo (*Hard Travellin'*, 1961). What was unusual about his writing about nature and the countryside from this period is that the voice and content were all of a piece with his 'cultural' writing. It was staccato, drily funny, full of contemporary references and metaphors. He was the first reviewer to recognise the genius of J. A. Baker's *The Peregrine* (1964). He wrote a remarkable piece for the *Sunday Times* in 1969 about a topic which had barely been glimpsed before – the teeming wildlife of London, followed by the script for a BBC film on wild New York.

This strand in his writing was, for me, a revelation, a kind of greeting. I was struggling to find my own voice as a writer, and coming (as Ken had) from a suburban middle-class background, found no resonance with the conventional tone of 'country writing', which at that time was for the most part either cloyingly sentimental or deeply conservative. By contrast, Ken's work seemed uncompromisingly modern. He was ecologically literate, politically fearless, and seemed able to speak to, and for, the growing constituency of urban nature lovers.

Towards the end of the sixties, to find a little respite from the environmental destruction he saw all about him and from his own metropolitan edginess, he and his family decamped to a 400-year-old mill house in the depths of west Dorset. For the next three years he contributed an offbeat weekly column in the *Daily Mail* (later collected in book form as *In the Country*) about his life in this landscape of ancient chalk ramparts and remote winterbournes – 'the loveliest part of the United Kingdom', he believed, 'the last place left'.

I was working as an editor for Penguin Books, education division at the time, and having met Ken a few times over the years, wondered if he might contribute a 'topic book' on nature conservation and land use for secondary-school leavers. He agreed. The short book, *Fit to Live In: The Future of Britain's Countryside* (the only title he was ever to publish on conservation), came out in 1970. Two years later, when my own first book *Food for Free* was published, Ken offered to write a feature on it for the *Sunday Times*. He invited me down to Dorset to spend a working weekend at his idyllic mill house. On the Saturday we took a long walk over the chalk whaleback of Eggardon Hill, the *axis mundi* of his personal landscape. He showed me a barn owl, roosting in a derelict lime kiln. I gathered a basketful of various weeds and toadstools for a wild lunch. But what I most remember about our ramble was becoming aware of the full extent of Ken's pain, not just the visibly physical, jaw-tightening ache of his stump as he climbed the strip lynchets, but the extent to which it also embraced the travails of the landscape itself, as if he and they were part of the same body. Not long before, he had learned of the Forestry Commission's plans to destroy the core of the ancient oak wood-pasture on Powerstock Common, which lay below Eggardon; and of Berkley Petroleum's closet prospecting for oil in the region, supposedly protected as an Area of Outstanding Natural Beauty. For Ken, these two developments were an attack on what was now his home country, but seemed to him also symbolic of the erosion of all that was precious about rural England. They turned him from a concerned naturalist into a militant environmentalist.

The story of his battles over the three years before his death is the subject of David Wilkinson's revelatory book. It is both inspiring and shaming. For all our cynicism about modern political sleaze, it is somehow especially shocking to see it alive and malignant forty years ago: the corporate lobbying, the biddable councillors, the suppression of information, the contempt for community feeling. Ken, in the face of this, became a force of nature. For three years he turned his rage and powers of rhetoric into what in the end was a kind of kamikaze attack on all environmental wreckers.

Some of what he did is hard to credit. Imagine a high-profile current affairs presenter today taking a very visible and political role in a public protest. Ken had no such scruples in 1972. At a mock funeral for west Dorset, he stood on the tor of Welcome Hill like a modern King Arthur and, watched by millions of television viewers, castigated an entire cast of planners, placemen and corporate wheeler-dealers: 'What is happening here is a warning of what the next ten years could bring throughout Britain ...' Or the few glorious weeks in which he was able to turn his new environmental magazine programme, *Down to Earth*, into what was in essence an on-air pressure group. The programme lambasted agribusiness, industrial polluters, government connivance and complacency. Astonishingly it got away with it for a while. The turning point came when the programme's special award for 'environmental idiocy' – the Plastic Banana – was given to Peter Walker, Secretary of State for the Environment, for the creation of Spaghetti Junction. Then BBC's management were outraged. The programme had broken the Corporation's rules about impartiality. Allsop was additionally reprimanded for taking part in the televised protests in Dorset. *Down to Earth* was emasculated, and moved to a provincial studio without rostrum cameras and graphics facilities. Then it was taken off the air altogether, and in the higher echelons of the BBC Ken was marked down as a loose cannon.

For a while he was able to channel his fury in a column in the *Sunday Times*, where there was no requirement for impartiality. Again he made frontal assaults on the new 'barbarians'. His tract – there is no other word for it – on the activities and questionable ethics of the mining giant Rio Tinto Zinc in December 1972 is regarded as one of the most powerful and eloquent environmental polemics of the post-war years. But RTZ sent in their lawyers, and after months of wrangling, the *Sunday Times* issued an apology. Then they began to censor Ken's pieces.

Allsop, fifty-three years old and a depressive by nature, was now facing a double despair, experiencing not only the destruction of all he loved but the silencing of his public responses. At the end of May 1973, he was due to face, on successive days, interviews with the deputy editor of the *Sunday Times* and the Controller of BBC2. He was convinced that he was about to be sacked by both institutions. The weekend before these fateful meetings he had been watching red kites and buzzards in west Wales, and had confided to Betty that he wished he could have his remains left on a hilltop for the scavenging birds to dispose of. On 23 May in what sounds like an untypically calm mood, he took an overdose of sleeping pills. He left a blackly comic note, suggesting he had a far more important deadline than the ones he was missing.

Ken's story was heroic, but also tragic, because of the inevitable failure of his crusades. Diplomacy was a process entirely foreign to his temperament. It is tantalising to ponder on what he might have achieved, with his charisma and rhetorical power, if he had been prepared to compromise, or perhaps retire to academia (he loved his brief fellowship at Merton College, Oxford, where he was researching the links between fascism and 'Back to the Soil' movements in pre-war Europe) and write the big book on the environmental crisis he had often dreamed of. But such moderated engagements with the world were simply not his style. It would be easy but insulting to make a clinical rationalisation for his sheer bloody intransigence: the need to maintain an external enemy is a common survival tactic for those in constant physical or mental pain. Ken simply did not believe in compromise, which he regarded as part of the reason for the degradation of nature he saw all about him.

But he left two important legacies. Many, like me, discovered in his distinctive prose voice something which matched the energy and resilience of wild nature. And, for the brief time he blazed across the sky, as the most high-profile environmental campaigner in Britain, he helped energise the nascent Green movement towards a state where it began to make the difference Ken was not able to achieve by himself.

The Commons Man

First published as a tribute to Roger Deakin in BBC Wildlife, *2006*

I owe many kinds of personal thanks to Roger Deakin: for his optimism about the world, for his passion for Suffolk, for his simple generous friendship. I wouldn't be living in his own stretch of the Waveney Valley now, with its ramshackle string of commons and fens and furze patches, if he hadn't taken me under his wing during a low moment of my life.

All of us in the natural history world owe Roger a debt, not just for two luminous books, but for an immense network of enthusiasms and initiatives, of causes championed, corporations badgered, of essays, notes, letters, beautifully crafted radio meditations, elaborate green jokes. It was more than a network, it was a kind of imaginative ecosystem, bubbling with life and inventiveness, and its energy touched people way beyond Roger's own locality.

But he wasn't some fervent missionary for the wild. He simply lived out what he believed in, in a quiet but infectious way. Central to his beliefs was a love for the common, in all senses of the word. He was a co-founder of Common Ground, the charity devoted to the celebration of the commonplace in nature and landscape. He lived on the edge of the great common of Mellis, and fought passionately to preserve its openness and ordinariness –

sometimes, when necessary, against over-tidy conservationists. For him, commonness – the blessings of continuity and abundance – and commonality, the sharing of land between fellow humans and their fellow creatures – were inextricably entwined, and the most precious, the most rare, of qualities. We often exchanged little news bulletins about what was happening on our home patches, four miles apart. Seemingly mundane things, like the arrival of the cuckoo, the state of the local bees, clouds of all kinds on the horizon. Occasionally I'd become overexcited about some prodigy – a flush of early marsh orchids in a thistly field – and Roger would respond, with the merest hint of a raised eyebrow, with a eulogy to the common buttercup. Once, his corrective, his tribute to nature's connectivity, was the September calling of tawny owls, 'answering each other all round Mellis', he remarked to me on the phone, 'and I guess all the way to you in Roydon'. He was exceptionally well read and knew he was echoing the poet Edward Thomas's 'Adlestrop', with its vision of blackbirds singing, 'Farther and farther, all the birds / Of Oxfordshire and Gloucestershire'.

His commitment to his home countryside was uncompromising. He'd restored his sixteenth-century farmhouse with his own hands, and it was his *axis mundi*, his laboratory, his museum. Dwelling properly there – he would not have approved of any description as bossy as 'managing' it – was his major project, and took precedence over (though was intimately connected to) the writing. It was full of found stones, twigs from rivers, bizarre outdoor shoes and cobwebs whose sweeping into oblivion he regarded as an act of wanton destruction. Roger saw his spiders as more anciently entitled citizens of the farm than him, the newly settled guest. Outside was a madcap Arcadia: hay fields gradually recovering their cowslips, raggle-taggle hedgerows, ancient cars driven into bushes and left there to form pergolas for brambles and foxes, new woodlets planted in concentric circles, shepherds' huts in which he often wrote and slept, wild roses grown from seed gathered in the Middle East, which alas he never saw flower.

Walnut Tree Farm was Roger's home, but not his limits. One of the standing jokes among his friends was that he had apparently been everywhere, done everything: run conservation projects in Africa, trekked with nomadic shepherds in the Greek mountains, trespassed in most of the desirable private wetlands in Britain. But the joke was on us: he *had* done these things. One of his last trips was to Kyrgyzstan, to live rough with the local tribes who camp out in the forest for the walnut and wild apple harvest, and he somehow smuggled home a sack of walnuts bigger than his luggage. He was quite fearless in his travels, believing that most beings were benign, and would respond to the philosophy of live and let live that he applied to all species.

I think Roger had an essentially comic – certainly untragic – view of creation. He felt he was part of a gleeful conspiracy of wild beings – spiders, badgers, climbing roses (which he used to cover anything he didn't like, rather than destroy it) – against the forces of oppression and order. His sense of humour was extraordinary. When Polly and I came to live near Diss, he sent us a welcome card, on which were mounted two empty honesty seed heads, and the caption: '*Lunaria annua* var Diss. VN: Diss Honesty. Originally rare and locally dist. Now widespread. Identify by: missing seeds.' His mischievous insect fantasy was to make an electronic organ based on the fact that crickets vary the pitch of their stridulations according to temperature. The crickets, encased in glass tubes, would be warmed and cooled by a keyboard control. In reality he wouldn't hurt a fly. But none of this was simply whimsy. Roger saw playfulness as a common ground between humans and nature, one of the places the poetic mammal could find a harmonious niche.

His travels in central Asia were for his book *Wildwood; A Journey through Trees,* an account of meandering through the woods of three continents. It was almost finished when he fell ill, and his editor was able to prepare the book for publication before he died. It will come out – a wry masterpiece on the entanglements of the human and the natural – in June next year [published 2007]. Till then, we have *Waterlog* (1999), his now famous account of swimming through Britain. *Waterlog* is partly about access to water, but also about immersion. To go wild swimming is to join the most fundamental and fluent natural entity on the planet, and to understand one's place in it. 'The next day I met an otter in the Waveney. I swam round a bend in my favourite river in Suffolk and there it was, sunning itself on a floating log near the reed-bed. I would have valued a moment face to face, but it was too quick for that. It slipped into the water on the instant, the big paddle tail following through with such stealth that it left hardly a ripple. But I saw its white bib and the unmistakable bulk of the animal and I knew I had intruded into its territory.'

Hidden Dips

First published as an essay for Edward Chell, Soft Estate, *2013*

> *I wanted your soft verges but*
> *You gave me the hard shoulder.*
> — *Roger McGough*

Junction 6, M6. It's not quite the portal to some indigenous Route 66, the first frame of a British road movie. But something like this was at the back of the minds of those who conceived the Gravelly Hill interchange in the late 1950s. What is now indelibly and affectionately tagged Spaghetti Junction was intended to be the futurist centre of Britain's expanding road network. At its opening in May 1972, the environment secretary Peter Walker thought we had arrived at 'the most exciting day in the history of the road system in this country … We are opening the motorway hub of Britain.' Eighteen elevated roads would intertwine in a Turk's-head knot just north-east of Birmingham, and provide the 'missing link' between the motorways of the north and south. In fact its extravagance, its constant looping back on itself, is as much a result of a parochial desire to tie together all of Birmingham's major roads as of any outgoing ambition. The great motorway hub is also just a concrete village green.

Experienced at different levels, this ambivalence – an essential feature of the topography of motorways – is even more striking. From the air, Spaghetti Junction looks like a gigantic piece of land art, a celebration of that central trope of nature, curvaceous connectivity. Driving on it, you have the illusion (and, first-timing, the panic) of not being able to find your way off. You're in a closed system, a Mobius strip, that mathematical conceit which consists of a flat circular strip of material with a single twist, so that it has only a single surface. The fact that, spiralling round, you have glimpses of another kind of geography below the asphalt carriageway – flashes of canal water, tangled green thickets, *footpaths* for goodness' sake – only heightens the sense that the motorway is a self-contained world. In its early years this dislocation encouraged all kinds of urban myths about what actually lay beneath. There were memories of the time MP Frank Field had suggested that 'difficult' families should be housed in vandal-proof steel containers under motorway flyovers. There were rumours that the council had created a gravel beach by one of the canals, with artificial sunlight fuelled directly from the adjacent National Grid power lines. More likely than either of these surreal possibilities was the realisation of the fantasy at the heart of J. G. Ballard's novel *Concrete Island*, in which a motorist crashes through the barrier of an urban interchange and is stranded for weeks in the occult triangle of waste ground between the roads.

What actually lies under Junction 6 is another kind of spaghetti, a tangle of converging canals and elevated tracks, braided with reeds and housing a colony of Britain's largest spider. The experience of the motorway landscape is continually jumping between these different levels: the hermetically sealed-off vista, the backdrop; and the tantalisingly unapproachable close-up, the foreground, sped by as soon as you've glimpsed it.

In the mid 1980s I was invited to devise and write a Motorway Nature Trail for the M6. The oddly paired sponsors were Gulf Oil and the government advisory body, the Nature Conservancy Council (NCC). I'd sat on the council since 1982 with a loosely defined responsibility to speak on behalf of the ordinary nature-loving public and about effective ways of communicating conservation messages. It was Gulf Oil who put the original idea for the M6 trail to the NCC, and it was a relief to bemused council members that they could shunt the project into my baggy populist portfolio. And quite soon I had been dragooned into writing it as well. To this day I'm unsure whether I should feel embarrassed about this, and whether I should have declined on ethical grounds. It happened before carbon emissions became the critical issue they are today, but I was still giving my backing to a resource-guzzling and polluting oil-based culture, the heart of Mrs Thatcher's chilling vision of Britain's identity as 'a great car economy'.

What persuaded me and the NCC to become involved was a line of argument that I think still holds good – just. If long-distance car travel is to be a fact of life, then it's better done on specially designed motorways than on congested trunk roads that are the legacy of a quite different style of motoring. And if the roads are there (regardless of what environmental havoc their creation entailed) why not make travelling on them a painless introduction to the landscape and ecology of the regions they pass through? Gulf's subtitle for the booklet (available free from all Gulf Oil service stations) was 'A new educational conservation guide for all the family'. But reading it thirty years on – and then following its route again – I'm slightly shocked that nobody felt it worth emphasising that the trail was principally intended for the *passengers* in the family. The programme of peripheral observation it outlines, downloaded into a driver's brain, could be as distracting as a mobile phone conversation. When I was doing the research for the booklet with my friend Philip Oswald, the scholarly director of NCC publications, we were driven in a chauffeured limousine. Travelling north Philip and I debated and made notes on the motorway's western embankments. We stayed overnight near the Scottish border, then travelled south the next day annotating the eastern verges.

Travelling cross-country on a motorway is closer to journeying by train than pootling by car on B-roads. You are on a ride whose speed and direction and circumstances are mostly out of your hands. Dawdling, shortcuts, spur-of-the-moment stops, backtracking and side-shooting are either illegal or impossible. The country beyond the road unrolls like film, a fast-frame precis of the landscape. What I tried to do in my M6 trail was to exploit this compression, and highlight the visible shifts in geology and vegetation as you travel between south and north. The industrial wastelands of the Midlands, tricked out with washes of rosebay, morphing into the heather-clad slopes of Cannock Chase; the intimate wooded ravines of Lancashire – called cloughs locally – giving way to the first dramatic upland vistas, the glaciated fells and dales of Yorkshire; sudden white outcrops of limestone by the edge of the road and then sandstone folds in a spectacular cutting near the gorge of the River Lune; granite erratics on Shap Fell near Junction 39, great boulders barged here by glaciers from the Lake District; the mountainous end of the road which stretches to Junction 44. A cross section of England's landscapes paraphrased in three hours' driving.

The same kind of highlighting is a side effect and occasionally a deliberate aim of the highway architects' desire to make motorways interesting and attractive, and therefore, hopefully, safer and less tiring. A motorway's route is determined largely by function and cost, but wherever practicable the carriageways are routed close to dramatic edge features – river gorges, rock outcrops, ancient woods. The tilt of curves and the positioning of roadside tree clumps are fine-tuned to open out panoramas of the landscape beyond, or screen out putative eyesores: subjective aesthetic judgements are as evident here as in any piece of landscaping. What is projected at the cocooned viewer, rushing past, is a virtual experience – but of a real landscape. The transparency of the car's windows which frame the scene, the interludes where embankment shades into field, give a sense of intimacy with what lies beyond the road, of the countryside as the motorway's 'off-scape'. The sculpted verge becomes, in places, the modern equivalent of the ha-ha. Iain Sinclair, in his epic circumnavigation of the M25, *London Orbital*, called the purlieus of the motorways 'the last great public parks' (though 'safari parks' might have been more apt, given the compulsory enclosure of the public in their mobile viewing booths).

I didn't twig how close this parkland analogy was until years later, when I began to look at the landscape theories of the Picturesque movement. Put simplistically, followers of the Picturesque judged landscapes by how well they would 'fit' as a picture. How their elements were balanced in terms of variety, structure, connectivity. William Gilpin, father figure of the movement, frequently applied these criteria to the landscape of roads. In *On Picturesque Travel* (1792) he wrote of how the experience of travelling 'the great military road' between Newcastle and Carlisle was improved by the vistas it gave of 'interchangeable' tracts of heather moor and pasture. He stressed the importance of road-edge as foreground, and the superiority of a winding road over a straight, 'which is, on the whole, much less pleasing, than a road playing before us in two or three large sweeps, which would at least have had variety to recommend it' – a sentiment which could have come straight from the rulebooks of modern road designers.

But Gilpin was more summer-holiday sketcher than aesthetic philosopher, and his recommendations that views could be improved by the careful placing of cosmetic ornaments – a group of cows, tumbledown hovels, ragamuffin children – is one of the things that got the Picturesque a bad name, as a kind of politically dubious dilettantism. The movement's real intellectual voice was Uvedale Price, and he saw visibly diverse landscapes as both agents of social integration, and demonstrations of cultural history and ecological variety. He deplored the authoritarian order of Capability's Brown's estates – 'drilled for parade ... like compact bodies of soldiers' – which he thought capable of driving a wedge between landowner and community. (Cf. on a motorway, the analogous potential for alienation between motorists on the one hand, and the road designers and notional owners of the roadside scenery on the other.)

Price developed a perspective where he began to see that the integrity of a picture and of a landscape were analogous, and that the act of *seeing* could be a route to ecological wisdom and Romantic understanding. 'Feeling through the eyes', as the historian of the Picturesque, Christopher Hussey, memorably expressed it. In his *Essay on the Picturesque* (1795), Price writes of the contribution of old trees to the roadscape on his Herefordshire estate: 'there is often a sort of spirit and animation in the manner in which old neglected pollards stretch out their limbs quite across these hollow roads, in every wild and irregular direction; on some, the large knots and protuberances add to the ruggedness of their twisted trunks; in others, the deep hollow of the inside, the mosses on the bark, the rich yellow of the touchwood, with the blackness of the more decayed substance, afford such a variety of tints, of brilliant and mellow lights ...' Later he laments the simplification, the de-wilding, of another lane, where a huge beech had stretched its roots over a bank: 'the sheep also had made their sidelong paths to this spot and often lay in the little compartments between the roots'. One day he'd found a gang of labourers covering the whole root system and its sheep beds with mould, and laying it 'as smooth from top to bottom as a mason could have done with a trowel'. Against this levelling he lays out in his argument all the elegant chaos of the natural: animation, neglect, wildness, irregularity, ruggedness, intimate detail. Biological diversity creating aesthetically fascinating visual structures. Price here is setting up a counterpoint not just between spirited green growth and what we might call 'brown stasis', but between foreground detail and background scenery, the resolution of which is easy enough 'on the ground' so to speak, but an elusive grail for the motorway designer trying to design naturalistic views for mobile travellers.

But it does happen. Nature's irrepressible inventiveness (Price's wild, stretching limbs) invades the tidy plans of the highway architects, who mostly welcome it (though many motorists don't). The un-treed stretches of motorway verges, apart from occasional scrub clearance and mowing for safety purposes, are pretty much allowed to develop as they will, and have turned into a great estate of feral grassland. A botanical survey of the M1 verges in 1970, ten years after it opened, found 185 species of plant had arrived of their own accord. In the following year a colony of native columbines was destroyed during work on the extension of the M1 in Nottinghamshire. Fortunately the county naturalists' trust had the foresight to collect seeds before the bulldozers arrived, and, with some poetic justice, got permission to grow them in the grounds of a motorway maintenance depot. Later; the mature plants were transplanted to the motorway verges as close to the original site as possible. They are still there.

These intensely local details are, of course, invisible at 70 mph, but they matter, because the organisms whose habitats the motorway has invaded are legitimate road users too. And enough have grown into significant features to become part of the foreground of the motorway landscape. Huge drifts of roadside cowslips blanket the chalky stretches of the M4. Thickets of those transatlantic settlers, Michaelmas daisy and goldenrod, have established themselves on motorway verges in the suburbs more harmoniously than anywhere else. Authentically wild daffodils – the very same species that Wordsworth rhapsodised about – splash the Ross spur of the M20 with two tones of yellow in late March. Nor is it just flowers that provide foreground 'animation'. Red kites are an almost dangerously distracting delight above the M40 through the Chilterns. Buzzards moping on verge-side posts are now commoner than kestrels, the iconic bird of the motorway verge in the last millennium. Rooks forage along the edges of the hard shoulder, where traffic vibration has driven worms to the surface; and there is an intriguing correlation between new rookeries and motorway service stations, driven perhaps by the proximity of more convenient food.

But it is a plant that provides the perfect parable of the motorways' wild landscapes. It knits the traveller's need for colour and structure at speed, and the opportunism of nature when faced with new, anthropocentric habitats. In the mid 1980s, colonies of a scarce coastal plant, Danish scurvy-grass, began to appear along the edges of motorways and major roads, especially in the prophetically named 'central reservations' It's a modest member of the cabbage family, but in late March these roadside colonies become so dense in places they resemble a layer of deep and persistent frost. When I did a rough paper survey in 1996 there were already concentrations along the M4, M5, M6 and M56, as well as a host of A-roads. Today there is scarcely a large road without it. The only inland region it has failed to colonise is Ireland, despite the species being native on Irish shores. What is different about Irish roads is that no salt is added to the winter grit. There are many reasons for the plant's spread throughout the UK road system – the turbu-lent slipstream of traffic, for instance, whirling the seeds along; the similarity between its native strandline habitat and the stone edges of the road. But there is little doubt that the major factor is the saltiness of the modern road – that shoreline tang sprayed from gritting lorries even in the landlocked heart of Britain.

Seen close to, Danish scurvy-grass is an undistinguished plant. Streamed by at speed, it is a ribbon of dazzling white at the motorway's grey edge, a traveller's joy. I call it wayfrost.

The Battle of the Bogs

First published in the Sunday Times, *1987*

What came to be known as the Battle of the Bogs in 1987 was a border conflict of almost Byzantine complexity. At its heart was the Flow Country, 1,500 square miles of wild peatland in Caithness and Sutherland, home to tens of thousands of wading birds and a scattered community of crofters. Ranged against them were the forces of commercial forestry, attracted here, as to many upland areas, by the expectations of cheap and apparently useless land, and hoping to create what would be the biggest single forestry estate in Britain.

It was the latest chapter in a familiar and long-simmering squabble about afforestation in Scotland. But the Flow Country had provided some extra and highly combustible ingredients. Distinguished conservationists were arguing that it was a wilderness of international importance, on a par with Serengeti and Amazonia. Landed Scottish interests countered that the whole landscape was a 'foul, bankrupt land ... a denuded wilderness through man's intrusion'. There were accusations of absentee landlordism and bureaucratic English meddling in Scottish affairs, whiffs of tax-avoidance loopholes and dreams of a job bonanza. When the press discovered that there were celebrities involved as well, that show-business figures such as Terry Wogan and

Cliff Richard owned blocks of 'investment forest' in the Flows, the argument became a hotly debated national issue.

It was a bemusing confrontation, on a scale more usual in land feuds a continent or two away than in our small and already intensively settled islands. But to some onlookers (and some of the participants) the Flows *were* as foreign as Brazil or Antarctica, a sodden, midge-infested quagmire whose cultural and biological values were inscrutable.

When I travelled up in late July it was unknown territory for me too, the furthest north I had been in the British Isles, and I wondered if it would overcome a long-standing uncomfortableness with bleak uplands. I was based in Golspie, a pleasant golfing and fishing resort on the east coast. It was full of boutiques and estate agents, and when I explored it on my first evening, with the swifts racing between the chic cottage conversions, it was hard to believe that this was the gateway to an expanse of primeval bogland the size of Lancashire.

Yet the region has been no stranger to land battles. On a ridge overlooking the town is a statue of the first Duke of Sutherland, clad in an imperious red sandstone robe. Early in the nineteenth century the English-born duke had been responsible for the notorious Sutherland Clearances. Between 1814 and 1820 he evicted a third of the population of the county from their homes (by burning them down in many cases) to make way for sheep ranching and shooting, and to provide cheap labour for his factories on the coast. His agent's defence of the purge has an ironic ring in the current circumstances. Their intention, he said, was 'to render this mountainous district contributory as far as possible to the general wealth and industry of the country, and in the manner most suitable to its situation and peculiar circumstances'. The Clearances are an ineradicable part of Highland folk memory, a symbol of all misappropriations of the land, and it was a measure of how deep feelings were running here that each side was accusing the other of initiating 'the New Clearances'.

I went back to my hotel and tried to make some sense of the tangled history of the affair. There did at least seem to be a ground base of scientific fact. What was at stake was the world's largest remaining concentration of blanket bog, a series of rain-drenched plateaux and pools with one of the richest collections of breeding birds in Europe: waders such as greenshank, dunlin, golden plover; loch haunters like red-throated divers – birds whose survival depended on expanses of remote, wet wasteland.

'Blanket bog' is a remarkable enough thing in itself. It is a living skin of sphagnum mosses, an intricate carpet of plants, unrooted except for their own mutual entanglements. Sphagnum is honeycombed with capillary tubes and twice as absorbent as cotton wool. In the very high rainfall of the region it becomes permanently waterlogged and swells sufficiently to blanket areas of moss that are dying back. This moribund sphagnum becomes part of the underlying layer of peat, and the live and dead moss form a moist ground base for other plants.

Yet the Flows has never been an uninhabited wilderness. For thousands of years it has also been the habitat of marginal farmers, who have grazed cattle, cut peat on the bogs, and raised crops in the valleys in ways that are entirely compatible with the bird life. Since the Clearances they have borne the brunt of the land-use changes.

Trees, until recently, played no significant part in the crofting life. They had not grown here naturally since prehistoric times, and the unrelenting winds and high rainfall made it difficult to grow them in plantations, except in sheltered valleys. The Forestry Commission, as a result, was a latecomer to the region, and most private forestry companies would not touch it. But in 1979 the Perth-based Fountain Forestry noticed a happy coincidence of business opportunity and technological know-how. Land prices were low (as little as £100 a hectare in some places), government planting grants were favourable, and machines were at last available that could work in even the most intractable bogs. Fountain began buying up estates in the Flows and selling them off to investors (mostly southerners: 72 of the 76 listed in the Scottish land register that year had English addresses). By the beginning of 1987 it had acquired 40,000 hectares and had earned its investors more than £12 million in grants and tax exemptions.

Fountain's expansion into the Flows coincided with an upsurge of interest in the cultural and biological interest of so-called wastelands. Since the early eighties, the Nature Conservancy Council (NCC) and the Royal Society for the Protection of Birds (RSPB) had been noticing drastic changes in the ecology of the afforested areas of the peatland, whose most conspicuous effect was a plunge in the breeding populations of wading birds. The NCC, as the government's statutory adviser on conservation, obtained an agreement that from February 1987 the Forestry Commission must refer to it all requests for planting grants in the Flows.

Things had come to a head a couple of weeks before, when the NCC published a report giving its scientific evidence on the threat to bird life, and making a plea for a moratorium on further planting over all the remaining 400,000 hectares of peatland. The report was met with almost universal hostility from the Scottish establishment. The Highlands and Islands Development Board said that it had been drawn up 'without consultation or regard for the delicate economic and social fabric of the northern Highlands'; Robert McClennan, the SDP MP for Caithness and Sutherland, described its scientific conclusions as 'preposterous' and predicted its proposals would lead to the loss of '2,000 jobs in the long term'.

The forestry lobby accused the conservationists of valuing birds above people and of meddling in affairs beyond NCC's remit. Conservationists retorted that it was not them but the alien conifers that were driving out – at public expense – both indigenous crofters and new tourists. Each side (it is the shadow of the Clearances again) accused the other of 'sterilising' the Flows, showing, to anyone who needed convincing, what a flexible and subjective concept 'productivity' is when it comes to land.

It was not just the scale and venom of the quarrel that were exceptional. Beyond the revelation of show-business stars' investment portfolios and the intricacies of government forestry policy, it was stirring ancient and unresolved questions about what we value land for. Was economic usefulness the only indicator? Did wasteland invariably mean wasted land? How could its value as a global asset be compared with its worth as a national raw material?

Later that evening there was music in the hotel bar. It was advertised as a ceilidh, but it was really a local talent contest for the tourists, an anodyne mixture of country and western and Kenneth McKellar. I recalled another musically tinged evening years before in Fort William, the frontier town between the Highlands and Lowlands. That night I had watched two pipers in full regalia sitting in the corner of a bar, playing laments while tears rolled down their cheeks, with their wives sipping gin by their side. The Flows felt like frontier country too, and with all the feuds, prospectors and bounty hunters it didn't seem too extravagant to view the galloping advance of 'the forestry' as a Great Timber Rush.

Next day I drove out into the Flow Country with my guide, Lesley Crenna, local officer for the NCC. Two features of the unplanted peat strike you when you drive in from the hillier land to the south. One is the sense of immense space, of a gently undulating flatness in which there are no straight lines, no harsh colours and, unless you look for them, very few foregrounds. In clear weather you can often see the peaks of Sky Fea and Genie Fea in Orkney, forty miles to the north. The other is what I can only describe as a kind of plasticity. The swell of the peat hummocks, the honey-coloured tussocks of sedge, the dark pools – all seem, if you stare at them long enough, to be actually on the move. Perhaps they are. The whole system is full of mobile water, which regularly seems to overload the sphagnum 'sponge' and spill out to form pools – the dubh lochans. There are many thousands of these, ranged in places like ladders, elsewhere in concentric arcs. From above they have the look of pool clusters on a saltmarsh at low tide, or the pitting in limestone rocks. Against this curving, tremulous landscape the severe stripes and rectilinear furrows of the new plantings seemed misplaced, and bizarre enough to have been created by extraterrestrials.

The Flows are really a kind of tundra. During the short, intense subarctic summers they buzz with life, with myriads of insects and wading birds. Four thousand pairs of dunlin (35 per cent of the European population) nest close to the pools, where they winkle out insects with their toothpick bills. So do the same number of golden plovers, and smaller numbers of greenshank, sandpipers and snipe. There are huge populations of meadow pipits and skylarks and, feeding on these and the waders and abundant voles, hen harriers and short-eared owls, with peregrines and eagles hunting out of the high ground adjacent to the bog.

I talked to Roy Dennis, director of RSPB operations in the Highlands and a crofter himself, and he admitted that the richness of the bird life is only just being appreciated, after years of survey work. It might have saved a good deal of misunderstanding if this information had been available earlier, but the remoteness and sheer extent of the land had made surveying a slow, laborious business, and no match for the speed of the new machines. Roy pointed at a map on the wall and said wistfully: 'The land is so flat you could start a bulldozer in Wick and drive it straight to Bettyhill.'

And that, more or less, is what had happened. Out of 65,000 hectares owned by Fountain and the Forestry Commission, roughly half had already been drilled into order with a gusto that would have done credit to seventeenth-century Dutch fen drainers. Giant excavators had dragged out drainage ditches in the peat, eight feet deep in places, and raised cultivation ridges for the rows of spruce and lodgepole pine seedlings. There were aerial spraying programmes of fertiliser and pesticide, and hundreds of miles of deer fencing had been strung across the moors.

Extraction roads had been built whose width was often double that of the local highways.

But as the trade association Timber Growers UK assured me, there was a plan behind it all. The original intention was to create an integrated forest, big enough to be economically self-sustaining and to support at least two new sawmills. One of the keys was the Lairg–Wick–Thurso railway. The new plantations had been sited as close as possible to this line, to minimise transport costs when felling came round. This, TGUK stressed, was why the industry was so anxious to continue its planting programme up to the original target of about 100,000 hectares. Only then could the planned economies of scale be realised.

Looking at the map I could see some logic in this. The railway meandered past many of the plantations – the older blocks round Loch Shin; Strathy Forest, where a sizeable chunk of a National Nature Reserve had been accidentally ploughed up; Wogan's woods near Broubster. But the line was economically precarious, the sawmills still a figment of an economist's imagination, and the whole enterprise looked less like an unfolding plan than an exercise in opportunism.

Even scientific fact has come to be regarded as negotiable currency up here. The most frequent demand I heard from Scottish authorities was for a 'court of appeal' against the hated, English-based NCC. Not just against its proposals (which were only recommendations) but against the evidence, which it was felt should be open to compromise. It is a relief that nobody had any illusions about ecology being an exact science, but disturbing that the age of a tree, say, or the nesting territory of a bird should be regarded as open to something like plea-bargaining.

The debate about bird populations was rife with such nimble legalistic footwork. It was still being argued, for example, that plantations *increased* the number and diversity of breeding birds – which they do, though only of those species that are abundant in woods and gardens throughout Britain, and, temporarily, a few birds of prey. The specialist birds of the wet peatland are completely unable to adapt. But, the foresters argued, surely they could be more sociable, 'bunch up a bit' in the gaps between the new plantations. As Michael Ashmole, director of Fountain Forestry, pronounced of the greenshank, the Flows' third commonest but still somewhat expansive wader: 'If a bird cannot survive on 650 acres, then it doesn't bloody well deserve to survive.'

A decade of intensive research had shown that the greenshanks do try to bunch up for a year or so. Then the stresses of competition and overpopulation start to show. The birds lose weight, lay sterile eggs, lose chicks by drowning in the drainage ditches or to crows patrolling out of the new plantations, and the inexorable decline into local extinction begins. Just as serious was the persistence of the argument that commercial forestry was simply restoring a landscape destroyed by earlier farmers. Although some of the region's prehistory is still obscure, fossil pollen and stump fragments show that the last time native woodland (chiefly birch, rowan, hazel and Scots pine) grew on the open plateaux was four thousand years ago. A decisive wettening of the climate after that meant that the young tree growth could not compete with the expansion of the moss on the peat, and woodland survived only in the better-drained valleys.

This is not to say that there hasn't been some glamorisation of history on the conservationist side. Lesley Crenna, a Highlander herself, told me that many locals were distressed to hear their homeland repeatedly described as 'the last wilderness' and compared with the Russian steppes or Alaska. Up here this wasn't seen as a compliment but as an insult to the work they and their ancestors had put into the land. It made them feel 'like savages'. Sometimes, Lesley told me, crofters burned off patches of the grass – not because it made much difference to the grazing, but to say 'I live here too'.

Even remote corners of the Flows showed the marks of hard subsistence farming. Lonely homesteads, a few small plots of barley and oats, thin channels cut in the peat to provide fresh water. These were difficult times for crofters, and a few had already sold out to Fountain. But what was alarming the Crofters Union more was a move towards speculative trading in crofting land. The holdings in this part of the Highlands were much bigger than those in the west, often several thousand acres in extent. Some had been purchased by outsiders, and there had been attempts to amalgamate and appropriate the common grazing so that this could be sold off for forestry. This was within the law of the 1976 Crofting Act, giving crofters the right to buy their hold-ings from the landlord for fifteen times their annual rental, but quite against its spirit.

We drove to Andrew Cumming's holding by the dubh lochans of Shielton, along roads lined with narrow peat diggings, each with the hand-written sign giving the owner's name. Andrew gave us a cheerful elevenses, but was gloomy about afforestation, which he likened quite explicitly to the Clearances. 'They burned us then, now they are blanketing us,' was his curt verdict.

Close to his homestead, now a feed store, men were fencing a new winter stock pen. Nearby was a more ancient stockade, knitted together out of waste metal and old farm machinery – a frugal way of recycling rubbish that no one would ever bother to collect this far out. But beyond the farm on the bog itself, the marks of humans and nature were less easily distinguished. It was too late for most of the breeding waders. But the dark crossbow shapes of Arctic skuas skimmed across the swaying plumes of cotton grass, adding to that persistent impression that the entire landscape was shifting.

Walking about in the oppressive humidity was a queasily disorientating experience. The sodden sphagnum rocked underfoot. It had an insubstantial, blubbery feel, like a jelly. Every square yard of it was different, a constantly reshuffled mix of staghorn lichens, sedges, and a dozen different species of sphagnum, speckled with the sticky scarlet jaws of insectivorous sundews. The pools were different too, encrusted with moss or full of bogbean, or edged with the golden stars of bog asphodel. Thin sheets of rain swirled in from the north-west blotting out first Sky Fea, and then the high Sutherland peaks to the west, until I hadn't the slightest idea in which direction I was facing. Several times that afternoon I had waves of inexplicable anxiety – something I'd often read of American writers experiencing in wild places, but hadn't expected only fifty miles from Inverness. The bog was criss-crossed by cryptic trails and thin ribbons of seepage water, as if the sphagnum had cracked. They may have been natural, but the bog's skin is so fragile that it can show traces of damage for years. One was the track of an otter, a darting, decisive run that plunged into the pool, parted the bogbean and slithered out through the sedges on the far side.

But another that Lesley showed me was a human trackway, the old mail route that was until quite recently the postman's path across the bog. She told me a story that showed better than any abstract definition where the Flows lay on the scale between wilderness and wildness. A few years ago a group of crofters were walking home along this trail in winter and became benighted. When they eventually reached a croft one of them was suffering badly from exposure. The others attempted to revive him by pressing hot stones against his feet, but overdid it, and succeeded in burning him. In the end he had to be carried off the bog on a stretcher – out along that old post road.

Time and again that day we saw what afforestation had done to this 'delicate economic and social fabric'. Croftlands had been flooded as a result of the diversion of water down drainage ditches. Silt from deep ploughing had been washed into salmon-spawning areas. Pesticide and fertiliser run-off was polluting watercourses and streams.

Even the economics of the operation began to look shaky, especially when you have, in addition, to consider the damage from windthrow on this exposed plateau, and the ravages of the pine beauty moth, an endemic pest which chews alien pines (but not natives) to ribbons. Some of the trees were growing better than had been painted by conservationists, but only because of the forestry equivalent of intensive care. The National Audit Office, costing out the business the previous year, was scathing in its criticism. The real increase in the value of the trees was little more than 1 per cent, which hardly justified the public subsidies being lavished upon them. The *principle* that tree planting should be eligible for public money is obviously a commendable one. It means for instance that local authorities can establish and support native woodlands as amenities for their ratepayers. But from any political point of view it is a preposterous waste of public money where no public good accrues.

Scanning the list of names and figures on the land register I wondered what conservation benefit had been seen in Fountain's operation by Timothy Colman, an ex-member of the Countryside Commission and NCC advisory committee. He owned 790 acres of afforested bog. Or whether Lady Porter, founder of the Westminster Against Reckless Spending campaign, felt there should be any cap on the £500,000 she was then legally entitled to in planting grants and tax relief.

Even the generation of jobs in the forestry sector looked like part of a prospector's fantasy. Fountain employed some sixty people in Caithness and Sutherland. The much-publicised figure of 2,000 jobs turned out to be a projection for forty or fifty years hence, when the first trees would be due for felling and processing. Over the whole of the Highland region only 1,500 people were employed in the forestry and sawmilling industries, and a good deal of the work was done by outside-contract labour, sometimes from as far afield as Germany.

Yet there were already models of a different kind of development here, which respected the wildlife and crofting traditions of the peat plateau, but encouraged alternative agriculture and new small industries in the valleys. At Berriedale, north of Helmsdale, for example, there was a flourishing native woodland of birch, oak and rowan, a spring-water bottling plant and an experimental wind generator (the local authorities wanted this to be painted conifer green, but it remained a defiant airy cream above the open moors).

I travelled home from Sutherland the long way, and out of the train window I kept catching glimpses of the North Sea, itself under siege from overfishing and pollution, and a reminder that the exploitation of wild places is no local problem. I dipped into Barry Lopez's wonderful celebration of northern landscapes, *Arctic Dreams* (1986), to try and get a wider perspective on the Flows issue. 'Confronted by an unknown landscape,' he had written, 'what happens to our sense of wealth? What does it mean to grow rich?' I hoped we were becoming wise enough as a people not to regard land as waste or sterile just because it had no overriding economic use. The Flow Country was already rich beyond accounting. Perhaps only a lucky few would ever see its wading birds, as they flew down to Africa. But its subarctic wastes, teeming with birds on the long summer evenings, are one of the great landscapes of the imagination, an engine of life for the northern hemisphere. If the foresters are sceptical about this, it would do them no harm to wait a decade or two till the arguments have settled. The peatlands have no such choice. They are the product of thousands of years of evolution and could never be recreated in our time.

The research for The Common Ground *made an indelible impact on me, shaping my views and creating some lifelong friendships. But the incident that made the greatest impression occurred in the Alkham Valley in Kent in 1978. That March, the Secretary of State for the Environment confirmed a Tree Preservation Order (TPO) on a wood that was known to be lying flat on the ground. Sladden Wood was an ancient copse just inland from Dover, rich in nightingales and orchids, including Kent's special glory – the statuesque Lady orchid. It also had a network of footpaths and walks, and the affection of almost the whole local population. In 1977 the valley had passed into the ownership of a rapacious modern farmer whose zeal for improvement was well known from elsewhere in the county. He set about clearing hedges, shelter belts and small woods, and ploughing up ancient downland, with the clear intention of converting the whole site to arable. On 23 November he moved his bulldozers into Sladden Wood. Acting with remarkable speed Dover Council procured a TPO that same day, and attempted to serve it to the owner. The events of the next few hours had the character of a black farce. The council officer pursued his quarry, but was repeatedly blocked by freshly felled trees. In their haste to get the wood down before the order was formally delivered, the contractors dispensed with the niceties of tidy felling and simply pushed many of the trees over or snapped them off some feet above ground level. At the end of the day it looked as if a tank battle had taken place in Sladden.*

The owner, believing his action enabled him to say 'Wood? What wood?', mounted a legal objection to the TPO. But by the time the case came before the inquiry the following March the splintered stumps were already starting to put up new shoots. The inspector's verdict was historic. He confirmed the NCC's argument that a tree remains a tree until it is uprooted or killed, and that Sladden 'will return to a woodland appearance even without recutting of the stumps', and that the 'reasonable degree of public benefit which must be established before a TPO can be confirmed can be future benefit'. Against the popular wisdom that trees are destroyed by being cut down, the verdict established a profound ecological truth as well as a legal precedent – that a tree's life depends ultimately on the survival of its root system and the integrity of the ground it grows in.

I went back a few years later to open Sladden Wood as a Kent Wildlife Trust reserve. The site – now nicknamed the Horizontal Wood – had survived the woodwreck. And, as had been predicted by the NCC, the trees had regrown from the stumps and were now several metres tall. The Trust, sensitively in my opinion, had decided not to tidy up the fallen and broken trees but to leave them as evidence of a historical event, and of the fact that cut trees and horizontal lumber were not moribund at all, but a habitat for multitudes of organisms, and quite often capable of sprouting into new growth. Lady orchids bloomed under a bull-dozed tree that had turned into a natural arch.

I became fascinated by woods' complex histories and the ways they had been cherished by local communities. Also by their extraordinary powers of spontaneous recovery, something that was confirmed by the after-effects of the great storm of October 1987. I spent weeks touring the wrecked treelands of southern England that month, and then revisiting them during the spectacular regeneration of the following spring.

But by then I had a wood of my own, a shot at creating a modern common and a kind of laboratory, where I could witness these weighty debates – management versus natural development, private versus common use – acted out for real.

Oliver Rackham

First published in the Guardian, *2006 (Rackham died in 2015)*

Halfway through Oliver Rackham's commanding new book there's an engaging photograph of the author, 'cleaving oak into radial planks with a froe'. I have seen many snaps of Oliver in the field, but I doubt there is a more revealing one than this. He's standing in light snow, his face hidden from the camera, and drawing the mysterious froe through an oak plank tucked underneath his arm. A man for all seasons, hands-on, familiar with the arcane lingoes of both trees and humans, and completely absorbed by the minute details of the world.

I was lucky to be at Rackham's public debut, at a conference in 1973 ('The British Oak'). The gathering had been convened by the Botanical Society of the British Isles as a multidisciplinary forum 'on our most widespread and revered timber tree'. But a group of younger and more conservation-minded contributors had decided that the conference was an opportunity to launch an attack on the timber trade's ongoing assault on our national tree. Not only was commercial forestry destroying oakwoods in favour of conifers, it was mounting arguments that the oak might not be native in Britain, because it failed to regenerate naturally inside woods, and that plantations were its only hope of a future. Oliver was the opposition's secret weapon. He was a young and rather shy Cambridge don (he had one elbow of his buttoned-up jacket darned red, and the other green), and addressed the seemingly uncontroversial subject of 'The Oak Tree in Historic Times'. But his paper turned out to be a bombshell, a clinical demolition of foresters' paternalism and an awesomely evidenced account of the fact that, for most of human history, trees had been regarded as a self-renewing resource. He made it clear what kind of writer he was to become, exact and iconoclastic. He dismembered old myths, challenged supposed historical facts with his own first-hand evidence. He described how he'd measured all the main timbers in the original part of his college, Corpus Christi (there were 1,249, mostly from small trees about seven inches in diameter), and calculated how frequently such a building could have been created from the renewable oaks of an ordinary Cambridgeshire wood. He blew away the notion that felling trees destroyed woodland.

In the half-dozen books he has written since, he has revolutionised the understanding of historical ecology. In elegant and robust (and often witty) English, and with a historical intuition as strong as his scientific rigour, he has laid waste the conventional wisdom of commercial forestry, the ideologies of theoretical naturalists, the 'pseudo-histories' of historians. His simple – and to him sacrosanct – precept is that the final arbiter in all arguments about woodland must be the trees and woods themselves, in all their dynamic, mutable, particular detail. For Oliver woods aren't abstract entities ('the Forest') but individual symbiotic networks of beetles, carpenters, deer, land-thieves, lichens, pollards, surveyors, toadstools. Nothing is typical. Everything is bespoke, locally grown. I once drove him to east Suffolk from Cambridge, a journey that should have taken about an hour. It took three, as he insisted on showing me a wood with five-leaved herb Paris, and explaining the details of a landscape in which he seemed to know every individual hedge.

In the foreword to *Woodlands* he lays out his credo – that trees are not 'merely part of the theatre of landscape in which human history is played out, or the passive recipients of whatever destiny humanity foists on them … [they are] actors in the play', with multiple interactions with time, and with all other organisms, including people – then concludes, disarmingly, 'For good or ill, I have no particular theory to promote.' Well, if that is not a theory, or at least a manifesto, I do not know what is.

When this book was announced (honoured before it was published by being chosen as the 100th volume in Collins' *New Naturalist* series), many of his readers wondered whether this would simply be an eloquent retrospective. But it reveals many new aspects of Rackham. The parish botanist, grubbing about in the oxlip coppices of his native East Anglia, has, for a start, become an international explorer. In Australia he's witnessed the spectacle of eucalyptus infernos, and 'the nonchalant way in which the trees carry on growing afterwards'. Fire is *necessary* for woodland growth in many hot parts of the globe. In Japan, he treats the local architecture to a Corpus Christi deconstruction. The five-storeyed pagoda of the Horyu Temple, eighty foot tall, and 'dated from tree rings to 670 AD', is built entirely of the rot-resistant cypress, *hinoki*. Japanese wattle and daub, still used in rural housing, is identical to ours – except that the wattle is bamboo. Ancient patterns and practices, working themselves through intensely local ecologies. In Texas, he sees unplanted hedges developing spontaneously along the lines of barbed-wire ranch fences, spreading from clonal oak groves called 'motts' – one of which, he's delighted to discover, was described in precise detail in an eyewitness account of an 1831 gunfight between James Bowie and the Waco Indians.

More significantly perhaps, Rackham has become more confident as well as expansive. He confirms his own past judgements on matters where it's been proved right, and in all other areas – true to his investigative ideology – leaves open questions for us to grapple with. One of his lasting legacies is the idea of what he dubbed 'factoids', myths which have persisted so long without scientific challenge that they come to be accepted wisdom – like the belief that building the Tudor navy was responsible for the widespread destruction of English woodland.

Woodlands, on one level, is a popular introduction to his earlier work, and a justifiable celebration of its findings. The myth of the Great Wood of Caledon, which supposedly blanketed medieval Scotland in native pines, is buried, and Dutch elm disease is revealed to have been present in Britain five thousand years ago. But it insists that we recognise the limits of our knowledge, which are partly due to the protean qualities of woodland itself. Rackham outlines the ancient origins of our woodland but has no firm answer as to why it was not the dense, impenetrable forest of legend, but, as fossil pollen proves, full of open spaces. Were they there from the start, eaten out by wild grazing animals or the result of storms and floods? Or were Mesolithic peoples doing more woodland clearance than we give them credit for? He is persuasive about the long (until a century ago anyway) symbiosis between trees and humans, and that, contrary to popular belief, intense industrial use of woods didn't destroy them, but helped conserve them, as valuable resources. Hence the continuous woodiness of the West Midlands, the Weald and the Chilterns. But what caused oak to stop regenerating in oakwoods in the early twentieth century, what Rackham calls 'the Oak Change'? Possibly a decline in traditional management, which provided the open spaces and light which this species needs to start life in. Aggravated, quite likely, by an invasive leaf mildew, a comparatively recent immigrant from America. Rackham talks often about 'storm effects' and in particular about the ecological benefits of the 1987 hurricane, in its disordering of bland and uniform forestry lots. But I sense a metaphor here, too. A kind of storminess is what real woods and trees live with continuously. They are not human pets or servants. They are dynamic, autonomous, resilient, *different*.

If there is little in the book about conservation policy, it is because this kind of understanding, this respect for trees as living individuals, is a necessary prelude to conservation. And if he is too dismissive for some of us about the current enthusiasm for encouraging wrecked farmland to evolve into wild, 'self-willed' woodland, it is because he believes that intricate and irreplaceable systems of existing ancient woods are that bit more important.

Oliver Rackham is a Renaissance man. No modern ecological writer has demonstrated in such lucid prose that humans and woods are ancient partners. He has little truck with the self-centredness of 'new' nature writers, but we are all in his debt.

The Man who Planted Trees

First published as a foreword to The Man Who Planted Trees, *Harvill Secker edition, 2015*

There are two Jean Giono stories. One is the world-famous fable about a shepherd who transformed a desert into a forest. The other is the story of the creation of the fable. Both could be called 'The Man Who Planted Trees' as both are narratives of growth and redemption, driven by the dreams of an idealist.

In 1953, when he was already one of France's most distinguished novelists, Giono was invited by *Reader's Digest* magazine to write a short feature on 'The Most Extraordinary Character I Ever Met'. Giono loved commissions. They took his literary imagination into unexpected areas, and over the next few days he completed the 4,000-word portrait of the shepherd Elzéard Bouffier that would evolve into the text of *The Man Who Planted Trees*. The *Reader's Digest* seemed delighted at first, but a few weeks later wrote again, with the indignant accusation that Giono was a cheat, and that fabrications posing as factual documentaries were not the kind of thing their eminent organ published. Giono responded with what amounted to a libertarian shrug of the shoulders, withdrawing the text from the *Digest*, surrendering his rights in it, and allowing it to find its own way in the world. Within four years, by ways which now seem slightly mysterious, the book had picked up its spare but powerful title and had appeared in print in at least a dozen languages. By the early years of the twenty-first century it had become a miniaturist classic, one of the key texts of a burgeoning green consciousness whose credo is that, even in a damaged world, imaginative human action and natural growth, working together, can generate a kind of salvation.

The book's narrator (I'll call him Giono for the present as the text is written in the first person) first meets Elzéard Bouffier during a long tramp between the rivers Drôme and Durance in Haute Provence, just before the First World War. Giono describes it as an area of 'unparalleled desolation' – treeless, parched, cindered with ruined farms and resinous lavender. The surviving villages are inhabited by charcoal burners – a bleak profession, the incineration of the already burned. Giono's first sight of Bouffier is as 'a black figure standing upright in the distance. I took it for the trunk of a solitary tree.' They make a brief friendship, and Giono discovers the shepherd's real mission. While out looking after his flock he plants acorns in the dusty soil of the hills, making holes for them with a long iron rod. He'd set a hundred thousand in the past three years, and had already made a nursery of seedlings to expand his regenerative mission into beech, the signature tree of the Mediterranean uplands.

After the Great War, Giono uses his demob money to revisit Bouffier's motherland, curious about the fate of the shepherd and his project. The tree planter, now in his sixties, is still spry and busy (though he has swapped sheep for bees) and his oaks and beeches are as high as a man and cover a tract eleven kilometres across. For the next twenty years Giono rarely lets a year pass without visiting Bouffier. And as the forest grows so the landscape itself revives. 'There seemed to be a sort of chain reaction in all this creation,' Giono remarks. The dried-out streams begin to flow again. The crumbling villages are rebuilt, and crops begin to be sown again on the lower slopes of the mountains. In 1935 a government delegation arrives to inspect this legendary, yet oddly unknown (given it is now thirty square kilometres in extent) 'natural forest', and promptly puts it under state protection.

Its existence isn't threatened until the outbreak of the Second World War, when the oaks planted in 1910 were cut to distil for ethanol as fuel. Bouffier – now eighty – 'didn't even know about it'. Like some inexhaustible evangelist 'he was thirty kilometres away, going peacefully on with his task, ignoring the 1939 war just as he had ignored the war of 1914'. Giono last visits the shepherd at the close of the war in 1945, two years before he 'died peacefully in the hospice at Banon'. So transformed is the landscape, so alive with artisans and incomers as well as burgeoning woodland, that Giono only knows he is there from the familiar names of the villages on the new bus route. The good shepherd had brought 'forth out of the desert this land of Canaan [and] I can't help feeling the human condition in general is admirable, in spite of everything'. Through the agency of his trees, Bouffier had created not just an ecological transformation but a moral renewal, in the midst of one of the blackest half-centuries of human history.

In the years after its diffusion (it is hard to pinpoint a moment of 'first publication'), *The Man Who Planted Trees* became a major inspiration for the idea of tree planting as a route to reparation, a way of making amends for the damage we had dealt out to the natural world. And it was almost universally regarded by its readers across the globe as a true story based on a real character. Only the diligent, and maybe over-literal, editors at the *Reader's Digest* had gone to the bother of checking the deaths at the Banon hospice, and found that there was no Elzéard Bouffier among them or among the previously living, nor was there any mysteriously risen new forest near Vergons. But anyone who knew the ways of trees and the landscape of Haute Provence would have found other disjunctions and contradictions there in plain sight – though one is forced into fact-checkers' pedantry to set them down. For example: if the landscape was barren and treeless, where was Bouffier finding his acorns, and the charcoal burners their firewood? The fact is that *The Man Who Planted Trees* was never intended to be documentary journalism. It is a short story, a piece of imaginative fiction, and in the 1950s not even experimental in its occasional use of real-life places and names. Giono was tickled by the public's readiness to believe they were reading a 'true' story, and in an 'Afterword' his daughter Aline recounts a jape he played on a German publisher, concocting a convincing but entirely fake photograph of Elzéard.

Giono's intention had been to write a parable about the contrast between human destructiveness and creativity, and tree planting plays both a literal and symbolic role here. The story has its roots in his experience in the Great War. He had fought at Verdun, often called 'France's Ypres' because of the scale of its slaughter. Ypres was where the artist Paul Nash painted his searing pictures of the horror of the trenches, landscapes of 'unparalleled desolation', with leafless and limbless trees rising out of the mud like broken soldiers. After the war Giono became an ardent pacifist, and much of his early writing was devoted to arguing the case against militarism. As he grew as a writer, this blended with an almost magic-realist view of nature and the Provençal countryside, akin to that of Marcel Pagnol (who made films of three of his novels).

Giono was profoundly disillusioned by the outbreak of a second world war, but never renounced his pacifism (which led to unjustified accusations of collaboration and a spell in prison). At the time he received the *Reader's Digest* commission in 1953 he must have been reflecting poignantly on the events of the previous decade, which included the progressive depopulation of the Provençal countryside. That he should have chosen to write an upbeat fable, an allegory about a proper relation between humans and nature, isn't surprising. But he thought he'd created something slight, and wasn't prepared for the scale of the responses and misreadings. In 1957 he wrote a private letter to a civic official in Digne, saying 'Sorry to disappoint you, but Elzéard Bouffier is a fictional persona. [My] goal was to make trees likeable. Or more specifically make *planting trees* likeable.'

His legacy on both counts has been considerable and positive, but half a century on — and wiser perhaps about the lives of trees and the human capacity for self-deception — we should be as wary of too literal an interpretation of his goals as of the story itself. Recently, tree planting has too often become an easy escape clause, a panacea for environmental ills that would have been better prevented than compensated for. It is emotionally rewarding to dig a hole and put a sapling in it, but it is a much less reliable way of starting a tree's life than allowing it to spring itself where it chooses. The pressures that have artificially boosted tree planting as a grand solution are enormous. Charitable organisations need it, and the targets that go with it, to get their grants. Corporations (e.g. oil companies) find it a convenient and cheap forfeit for fossil-fuel abuse, and a useful PR gesture. It's been done often in the wrong places and with the wrong species, or where it would have been better to have let nature itself decide the planting programme. Giono has Elzéard say ironically to the state foresters, determined to protect his 'natural' forest, that 'this was the first time ever ... that a forest had sprung up of its own accord'. But of course forests do spring up everywhere of their own accord (and are repeatedly cut down, by conservationists among many others), and have done so for aeons. To forget this is to ignore nature's own capacity for reparation, and feed our hubris.

Perhaps the way to read Giono's book in the twenty-first century is to fully respect his imaginative intentions. The trees, like the story, are allegorical, symbols of reciprocity. We can gift them to the earth, but the earth, properly treated, also gifts them back to us.

A Wood over One's Head

Based on an essay published in Ground Work, *edited by*
Tim Dee, 2018

It was pinned to a beech tree in the lane like a poster announcing the circus was in town. 'WOOD FOR SALE'! Bold letters, no stated price. A path tumbled down into a dry valley then up again. There were statuesque beeches to the right, frayed ashes trapezing to the left. Through their trunks the haze of spent bluebells looked like dry ice. Who wouldn't be tempted? But the prospect made me giddy. What was I doing, dreaming of buying a wood when I didn't even own a house? My rational self was ready with some thoroughly sensible reasons. During the seventies and eighties my work with the Nature Conservancy Council and subsequent explorations and reading had made me enthralled by the mystique of ancient woodland, by its intricate interiors and cryptic ecology, and the knowledge that human communities had lived with it on intimate terms. It was still in trouble nationally at the hands of rampant development and commercial forestry. Might it be possible for me to have a patch of my own, 'save' it, and re-establish that intimacy?

But thirty years on I recognise that there were other kinds of neediness at work. I was almost forty years old and single. I no longer had a bolthole in East Anglia, and was finding the business of freelance writing increasingly lonely and precarious. A bosky retreat rather more ancient than myself would have been anyone's sensible prescription. I think I was at a point in my life when I wanted to make something more solid than books, to leave a legacy, maybe have a place where I could go feral among the trees again as I had as a child. What I hadn't anticipated was how the adventure of ownership would be inexorably entwined with the burden of control. For the next twenty years I felt I was in a permanent seminar on the interaction between nature and culture, debating how my romantic ambitions and the wood's own self-willed agenda might be reconciled.

One problem was that to dis-own a wood, so to speak, you first had to own it, and play the property game. That anyone could be said to possess a living entity that had quite likely been evolving since the end of the Ice Age has always seemed faintly absurd to me. But that is the way of the world. I had to man up and be prepared to deal with fences and planning laws and the insidious social pressure to *do* something. I did not accept this frame of mind with much grace, even at the very beginning. In the months I was negotiating with the agent, I mooned about in Hardings Wood (as it was called), hiding in the undergrowth when yet another bidder with a Norfolk jacket and a clipboard hove into view and made me wretched with gloom and frustration. In the end I sealed the deal by contacting the vendor directly, and getting emotional about my plans for the place. It was no skin off his nose, as I would have to pay the same rate as a forestry company. (I actually paid rather more, the vendor soon realising I was a non-commercial and possibly desperate customer.)

Objectively it was a more promising site for me than any forester. Less than a third of the sixteen-acre site was devoted to planted beeches, then about ninety years old, and of some commercial value. The rest of the wood was slung like a hammock over a dry valley a quarter of a mile from the village of Wigginton in the northern Chilterns. My research in old maps and deeds showed that woodland had existed on the site since at least the sixteenth century, and, the prodigious ground flora suggested, for a lot longer than that. The current tree cover was species-rich, with hornbeam and sessile oak and rowan, but most of it was thick, forty-year-old natural regrowth after the place had been gutted of timber during World War II. There were ash and cherry trees in the valley, birch and holly on the clay plateau, but they were all worthless as timber. And sometime during the 1960s the last owner, an absentee wood merchant excited by an ephemeral subsidy scheme offered by Bryant and May, had over-planted the whole site with matchwood poplars. They had become infected by honey fungus on the unsuitable clay soil and were dying on their feet.

The wood was dark from the dense regrowth, but full of shade-loving ferns and flowers. I wandered about among its strange and unremarked tuft-ings – beeches shaped like candelabra, dynastic badger setts, violet orchids glimmering in the deepest shade – and wondered what I could make of it. I was full of hubris then, expert in ignoring what I knew perfectly well, that in the long term the wood would survive and flourish even if I did absolutely nothing. But I was passionate to get engaged, to let in some light (in a more than literal sense), to make the place wilder, woodier, more beautiful in some ill-defined way, to feel I had enhanced the earth's growth a mite before I was added to the leaf litter myself.

The novelist E. M. Forster owned a wood in Sussex in the 1930s, and confessed that he too was troubled by this compulsion to intervene. Possessing Piney Copse, he wrote in an essay entitled 'My Wood', 'makes me feel heavy. Property does have this effect ... [it also] makes its owner feel that he ought to do something to it. Yet he isn't sure what. Restlessness comes over him, a vague sense that he has a personality to express – the same sense that, without any vagueness, leads the artist to an act of creation.' I never felt that kind of grand creative impulse towards Hardings, but there was no vague-ness about my fantasy of restoring one small plot of the Greenwood to the

People. I declared my ambition to establish a 'parish wood' by cutting the fence across the entrance.

And then, with the help of local friends and allies, I canvassed the village, asked for memories of the place, invited the inhabitants to come and help with the 'work', whatever that might turn out to be. The response was mixed, revealing the different declensions of possessiveness people can feel about places. A taxidermist, bizarrely, was the one soul to express outright hostility, protesting that activity in the wood might disturb his questionable trapping vigils. The master of the local foxhounds imagined that sharing a few whiskies might persuade me to open the wood to the hunt. A handful of villagers fretted that the phrase 'community wood' presaged coach parties from Islington. Not that they seemed engaged as a community themselves. When I talked about our scheme to the local primary school I was shocked to find that not a single child had ever slipped into the wood, despite its comparative closeness. No dens, no bike scrambles, no dirty games in the deep ditches. Things had been different half a century before. The oldest villagers could remember the 1930s, when the local woods were worked by cross-cut saw teams, and kids gathered up the debris in shopping bags as fuel for home. One woman showed us a basket plaited from willow and hazel by the Italian POWs who had looked after Hardings Wood during the war.

It was no great surprise that on our first working day locals were outnumbered by friends and fellow woodland enthusiasts. A score of us assembled one autumn Sunday with the beech leaves turning golden and the bracken tipped by frost, and realised we hadn't a clue what to do. This, so far as I knew, was the first attempt to create a community wood by private enterprise, so there were no precedents from the wider world. Nor was there much evidence of ancient management practices we might carry on, certainly no preening 'heritage' model to live up to. That morning all my nebulous visions about the place collapsed in the face of the stark prospect of making the first cut.

So, a little nervously, we talked. Those present shared their feelings about the place, and how they felt it might be enhanced. Tree identities and the merits of open space versus shade were debated. This was a few years before there was irrefutable proof that Britain's primeval woodland was an open savannah not a closed canopy forest. But I think we all intuitively felt the wood needed more open space, that this was a more 'natural', or at least more desirable state. We had met by a woodland pond, a very dark and overshadowed body of water, and clearing a glade around it so the sun could get in seemed uncontentious. Pardons were granted to characterful oaks and hollies whose case was argued before this impromptu forest court. We set to with chainsaws, the thinned trees being logged up with handsaws. At the end of the day we found we had a created a curious but pleasant miniature landscape, like a patch of savannah around a waterhole. Then we had a very sociable bonfire of the brash and toasted parasol mushrooms over the embers. (The end-of-the-working-day bonfire tradition didn't last long, despite its appeal. We realised how environmentally damaging the fires were, and thereafter left surplus wood to rot back into the ground.)

And that was pretty much how work proceeded over the next few years. It would be fanciful to call it on-site democracy, yet it wasn't some rigid pre-established management plan either. What gave our actions a kind of coherence was that we were all responding to the prompts the place gave us as much as to our personal visions. The footpaths (there were none at the start) were trodden out by people responding to animal tracks and natural contours and the magnetic pulls of viewpoints and big trees, or just taking obvious shortcuts. Everyone felt that letting in light was a priority, so thinning dense stands of ash and sycamore and poplar soon became routine. The populations of ground plants and warblers responded generously. Apart from the bluebells we had sheets of wood anemones, seven species of fern, three of orchid, and breeding blackcap and chiffchaff.

For the first time in half a century villagers came to enjoy the wood. I once chanced upon a couple of watercolourists seated among the bluebells in straw hats and looking much more a part of the place than I was. Moss was gathered for flower arrangements, and firewood by anybody who had helped with the work. A bat group clocked three species with their detectors. New customs were established too. On Ascension Day the children from the village school tramped over the fields, and in a clearing among the bluebells and freshly leafed beeches sang hymns to new life and the mysteries of transubstantiation. Then, released from the constraints of theology, they romped among the trees like fox cubs. In summer, after exams were over, there were sixth-form parties and woodland sleepovers.

Some part of my own dream of ushering a parish woodland into existence had happened. But with hindsight I realise I was driving the agenda much more than I admitted at the time. On one occasion I acted as an outright commercial forester. The big beeches which had been planted on a field adjacent to the wood at the start of the twentieth century were, on a forester's time sheet, ready for thinning. But of course they didn't *need* thinning. They would have survived as a group of trees perfectly well for at least another two centuries. Nonetheless they had been planted for timber, and thinning them would contribute to our philosophy of enlightenment. And, I reasoned defensively, we needed a small injection of revenue to pay for a pickup truck to fetch out cut wood ... I salved my conscience with the high-minded phrase 'continuity of intention'. So one November day I found myself progressing through the wood with a professional surveyor, deciding trees' destiny and the future interior landscape of this part of the wood with the help of cans of different coloured aerosol paint. My mentor would choose those trees that might eventually generate fine timber. I could pick those that were picturesque or promising dead-wood habitats, or just tickled my sentimental fancy. Between these, and marked with red dots, were the trees that could comprise an immediate moneymaking harvest, mostly close to the chosen ones, so that their removal would give these more space to flourish. I was pleased to learn that the whole process is known in forestry jargon as 'selection'. It sounded as if we were preparing a Tree Show for the Royal Academy.

I had a private relationship with the wood that was separate from guiding the community project. Sometimes I daydreamed about acquiring a traveller's caravan, and installing it in an inconspicuous corner. But the notion smacked too much of private appropriation. So mostly I just mooched about by myself. Walking the same tracks two or three times a week I could spot the tiniest changes, which ash seedlings had put on an inch in height, where windfall branches had dropped from, how our minimalist experiments with the light were affecting the rhythms of growth (not a lot). Although I had no interest in finding some transcendent meaning in the 'forest experience', I was touched by these intimate details, which gave me the sense of belonging to an exact and particular place. A corner where the perimeter path turned south and the bluebells and beech leaves always opened at the same time, was a tunnel of strange submarine light. A hornbeam half tumbled had formed a natural archway. I had favourite nests, where shade and slope had combined to create natural glades. Sitting in the centre of a huge clump of holly, or among a circle of wood anemones, I had the sense of being in a still centre, with the wood turning infinitely slowly around me. When I return to Hardings now, more than ten years after I moved away from the Chilterns, I find those two decades of intense memory are still there, and I'm able to see the long-term consequences of the fall of a particular birch tree on a precise day in January in the 1990s.

I paid obsessive attention to the wood's flora then, which was shifting partly because of the work we were doing, partly from its own cryptic cycles. Holly spread dramatically, causing a decline in our native daphne, spurge laurel. But primroses thrived on the edges of the new pathways. Wild garlic appeared, as if by magic. Sweet woodruff jumped out of the ancient wood and established itself in the new clearings in the beech plantation. One May, working on my hands and knees, I counted eighteen colour variants in the bluebells, from pure white and white stripes on pastel blue to dark indigo. I had volunteered to survey the wood for the Hertfordshire County Flora then under preparation, and a generalised ecstasy over the spring florilegium wouldn't do. So I learned to evaluate glumes and lemma in the woodland grasses (our single tuft of wood barley was a conspicuous bearded oddity among the starched anemones). I tracked the migrations of supposedly immobile ferns from their redoubts on the medieval woodbanks to nondescript addresses among the planted beeches. I realised that our exquisite wood vetch – sweet-pea-scented, liquorice-striped and the only colony in the county – depended for its survival on the disturbance we created. Our one major piece of hard landscaping was to have a track excavated from the valley up into the beech plantation so that timber could be tractored out. Wood vetch lined the edge of this track for the next few years After a decade Hardings had, relative to its size, the highest score in the county of what ecologists call 'ancient woodland indicator species'.

In winter the wood was a highly dynamic environment. The early nineties were an exceptionally rainy time and sometimes I would go there after a downpour and hear an unfamiliar sound, the lap and gurgle of water running over stone. The current of rainfall pouring down the lane had found an easy left-turn at the wood's entrance and had made off down the hill. It fingered its way into every track and rabbit run. It carved out runnels down to the bare flint, piling up dams of woodland flotsam – beechmast, dead leaves, a leavening of fine gravel – and laid out terraces with miniature pools and rapids. For a few hours most rainy days I had an upland stream on my patch.

The heaviest downpours sent flash floods rasping through the wood. Everywhere the water took shortcuts, dashing down badger trails, bursting through hedges from the fields next door. In the valley one day it built up to a torrent that tore clumps of fern out by the root. It snapped off a branch from a fallen tree and drove it forward like a snowplough, wiping a swathe of ground ten yards long clean of vegetation.

I sometimes tried to imagine what violations an eighteenth-century landscape gardener might have committed, unable to resist making permanent the seasonal streams in the valley. There had been moving water more recently. In the 1930s, when the Rothschild family owned the wood and many of the houses in the village, they'd laid sewage pipes from the cottages down the hill, to eventually gush their contents among the cherries. Fifty years on trees would still not grow in the eutrophic midden that resulted, and insisting that this unwholesome traffic was stopped was the first action I took after buying Hardings.

One other thing I did in the winter months was to try and find out more about Hardings Wood's history, and the history of the parish in which it was situated. There was a story here – the converse of ours, I suspected – about how the place had gradually ceased to be a community asset.

The earliest map I was able to trace, from 1766, showed Hardings as a compartment in a much bigger tract of woodland, a mile and a half long and a mile wide at its broadest point, which stretched all the way from Wigginton to the western edge of Berkhamsted. It abutted three hundreds acres of heathy commonland shared by Tring and Wigginton, which in the late 1600s, Henry Guy, Secretary to the Treasury, attempted to appropriate and turn into a private park. The locals took direct action. Daniel Defoe describes in his *Tour Through the Whole Island of Great Britain* how, finding Guy's agents already erecting fences, 'they rose upon him, pulled down his banks and forced up his pales ... and this they did several times, till he was forced to desist'.

By the time of the Tithe Commutation Map in 1841, two-thirds of the woodland had been cleared for agriculture, leaving just five island fragments, of which Hardings was one. A decade later, 'improvement' and large-scale agriculture pushed even harder against the parish landscape. The Lord of the Manor, together with all the major landowners, petitioned Parliament for the enclosure of the whole of Wigginton Common. They met in a local hotel. I found the records and memorabilia of the process of enclosure preserved in the Herts County Record Office, down to the hotel bills and cheque stubs for legal fees and the surveyors' little sketch maps (with comments pencilled on). The benefits of enclosure were set down as 'increased productiveness of the land, useful employment of labour ... and improvement of the morals and habits of the people'. On 7 August 1853, the twenty-four landowners agreed a total reorganisation of the parish's communication systems. A fortnight later posters went up all over the village, set in heavy block type, much like the 'For Sale' notice that had kick-started my own involvement in the place. Thirteen existing tracks and footpaths were to be completely closed off, and replaced by two straight surveyors' roads. One of them was a track which led from Wigginton village, down through Hardings Wood to join eventually with the main Tring–Berkhamsted road. No evidence that it had ever existed remained.

By the end of 1853 the enclosure was complete. The common had been ploughed and fenced, and the commoners and landless poor of Wigginton were granted by way of compensation a small recreation ground and five acres of allotment. This was pretty much par for the mid nineteenth century, but I still found it odd that the village put up no resistance, as it had so successfully two centuries before. Then, thanks to the local vicar, I learned of an extraordinary sequel. After the uprising to save the common in the seventeenth century, and maybe also because of its geographical position, nestled among woods in the hill country of the northern Chilterns, the village had acquired a reputation as a resort of lawlessness and immorality. 'Wicked Wigginton' was the tag it had acquired in the mid nineteenth century, as if it were some redneck settlement in Appalachia.

In 1854 a new curate, George Gaisford, was appointed, charged with writing a confidential report on the state of the village's soul and, if necessary, attempting to restore respectability: in essence he was the bishop's spook. His notes about the parish – which he called 'a *terra incognita* in the neighbourhood' – were kept secret in the diocesan archives for a century and a half, for fear they might offend descendants of the original families still living in the village. When the current vicar obtained permission for me to read Gaisford's report, it was not at all what I expected, and not only exonerated the inhabitants but made some questioning remarks about the enclosure. 'All the picturesque appearance of the place was gone, and perhaps the poetry. Post and rail fences were left and right all over the place. But I believed that the change would morally tend to the benefit of the people: they would be less rough, wild and uncivilised. I cannot judge whether this has happened: whether as some predicted "wicked Wigginton" would become "virtuous Wigginton". I cannot say that the people as I knew them deserved the former epithet, or differed much from other people of whom I have experience. But as I look back I am much surprised that they accepted the enclosure as patiently as they did, considering of how many rights the enclosure deprived them.'

One side effect of the enclosure and the closing of local trackways would likely have been an end to easy access to the wood, and to the gathering of fuelwood. I hope our new regime remedied that. But the parish history didn't provide any justification for our more fundamental interventions. I was perfectly aware that Hardings Wood would survive without human attention, and perhaps develop into something more spontaneously interesting than my ill-defined vision. It was part of conventional conservation wisdom at the time to argue that landscape, and possibly even 'nature' itself, was a human creation, and woodland a human artefact. Even as a philosophical conceit I've never got this. Our (and our predecessors') actions – cutting down trees, lopping branches, creating trackways – were acts of manicure, not creation. The wood doggedly insisted on retaining its identity as a wooded place, not some other kind of habitat, considering our interventions for a while and swatting them aside when they were inappropriate. The stumps of felled trees regrew reliably. Patches of open ground turned remorselessly into scrub, or young woodland. The essential contours of the ground remained unaltered by our superficial scratchings. It would have been possible to remove Hardings Wood from the map, but it would have involved the clearance of the trees, stumps and all, and annual cultivation of the ground in perpetuity. The 'nature' that grew there instead was not some human construct, but the solid results of aeons of evolution, of transactions between light and matter that we are still unable to replicate.

My personal contributions to the cosmetic look of the place were modest. Sometimes I would work there by myself in winter, sawing up logs from thinnings we'd generated on sociable weekend work parties. More often I would saunter round the place with a pair of heavy-duty loppers, doing just the sort of tinkering I despised elsewhere. I clipped the brambles round my favourite primrose patches, so they would make a better show. I pruned low-slung cherry branches to create arches over the tracks. I lopped sycamores that were shading ashes, and ashes shading young beeches, as if I had certain knowledge of the proper hierarchy in trees. As for the damned poplars, I was in the habit of looking for ones about to topple and nudging them into a position where they would take other poplar branches with them when they fell. And I remember the excuses I made to myself for this finicky stage management in a place I liked to think of as halfway wild. I was doing something no less natural than the local badgers or a tribe of bark beetles might do, making corners of the wood commodious for themselves. I reckoned I deserved a niche along with the rest of creation.

The curious thing is that I experienced not a mite of the tinkering instinct outside my wood. Everywhere else I was delighted by muddles, by obscured flowers, by views from, not views of. I think William Wordsworth was right that it is partly the licence given by ownership and power that encourages gratuitous intervention. In 1811 he had visited Foxley in Herefordshire, seat of the philosophical high priest of the Picturesque movement, Uvedale Price. Wordsworth admired the estate, but not the owner's habit of lopping branches to open up a view, or levering a few rocks into the splash of a stream. '[A] man little by little,' he chided, 'becomes so delicate and fastidious with respect to forms in scenery: where he has a power to exercise control over them and if they do not exactly please him, in all moods, and every point of view, his power becomes his law.' Foxley 'lacked the relish of humanity' of a 'country left more to itself'.

I moved to Norfolk in 2002, and with no wish to become an absentee land-lord sold Hardings Wood to a village trust – a sadness for me, but a boon for the wood, which now has a written constitution and a guarantee of open access indefinitely. When I go back I think I have been too hard on myself. The wood does have Wordsworth's 'relish of humanity', and an obstinate, self-willed temperament too. In response to what we did there during my tenancy, it's been generous. The footpaths we trod out have firmed up, but not dogmatically, and wander pleasantly wherever a fallen tree or a new badger digging intervenes. The gaps in the beech plantation where I ordered the executions have filled with spectacular natural regeneration, of more than ten species of tree. We did plant a few saplings, mostly gifts it would have been churlish to refuse, but their weedy growth has been a caution by the side of the self-sprung wildings. And new residents have arrived – wild daffodil, nesting buzzard, fallow deer from the big herd on the Ashridge estate, two miles away, and including on one occasion a pure white stag.

As for myself, I'm not entirely sure how those twenty years of tenancy affected me. Did that vague neediness that made me buy the place find some satisfaction? I think it did, but not at all in the way that might have been expected. I never felt that I was a 'steward', with some kind of paternal responsibility for the place. Everything that happened there, from the rampant spread of holly to the way the wood brushed off the effect of the two great storms of October 1987 and January 1989, challenged the dominant cultural assumption that the natural world is intrinsically fragile, in need of constant care to survive. If anything I felt the opposite, that the wood was looking after *me*. It proved to be a refuge of inspirational growth in the years I was helping to nurse my mother, invalided by Parkinson's disease. More selfishly I found it acted as an immense research archive, full of prompts when journalistic ideas dried up, or setting me off on enthralling hunts through woodland and botanical history.

Given what I now know about the natural world's resilience, I think I might have intervened less. Hardings was too small to benefit from the full surrender to natural processes now called 'rewilding'. But I wonder, for instance, what might have happened if we had left all those decaying poplars to die and rot where they stood, what dark melee of fungi and dead-wood beetles might have developed there. Yet I doubt I could have resisted becoming physically involved in some way. I don't think I – or any of the people who worked there – wanted any kind of 'dominion over nature'; just to regain a sense of being part of the place, and become modest forest creatures ourselves.

Still Lives

Developed from an essay originally published in Arboreal,
edited by Adrian Cooper, 2017

I keep it on my desk as a memento of where I grew up, and as a constant
warning against hubris. A turner had cut a section from a beech trunk and
opened it up into a bowl, and in the process unwittingly created a frozen
picture, a still life, of a woodland ecosystem. The bowl itself is a striking
object. It's massy, about eighteen inches across and finished with an exquisitely
silken lustre. A slight cracking in the wood has been tracked with a gilt inlay.
But its form is overshadowed by what is happening in the wood. Even the
hollow that defines its function seems part of another narrative, its gently
curving surfaces acting like a lens on the real business of this piece. Running
across the pale surface of the beech and diving into the pit is an extraordinary
lacework of dark lines. It has the look of a contour map of rocky terrain, or
maybe an exercise in woody tie-dye. The lines meander, fork, run close together
for a few inches. They outline ridges and plateaux and small, abrupt bluffs.

The patterning is called spalting, and is caused by what foresters would
call a disease. The chief organism responsible, the fungus *Kretzschmaria
deusta*, is a serious rotter, of beech especially. But it's an impeccably native

species with its own role in the variety of life, and is only really malign from a forester's point of view. The negotiation it performs with the beech, and with other individuals of its own kind, is more fairly seen as a complex dance of advance and retreat and territorial compromise. The fungus grows membranes of tissue between the woody cells of the tree, and attempts to expand along its water-conducting vessels. The tree responds by producing tannin and other fungus-repelling chemicals, and dark zones form in the marrow of the wood. The fungus argues back, building its own barriers of hard tissue as a defence against the tree's antibiotics. These make up some of the lines that edge the dark zones. More result when individual colonies of the fungus drift like fluids through the solid wood, come up against each other, and create mutually defensive barriers, often running in parallel. The tree becomes occupied by the fungus. If patches of wood begin to rot, the tree attempts to isolate them. If the fungi are frustrated they probe other vessels and fault lines. The end result isn't often a clear 'victory' for either side, but a kind of dynamic equilibrium, of which the interior of the trunk provides a cryptic fossil record. Then the trunk was felled, stopping this process in its tracks, and the turner opened it up, revealing a three-dimensional chart of its history that is profoundly beautiful and meaningful to human eyes, but which has no visual function whatever for the participants.

I ponder my bowl endlessly, and can take no simple lesson from it. It's both a literal and metaphorical representation of the dialogue between pattern and process that is at the heart of all living systems. The cutting of the tree and the turning of the wood, like all acts of 'management', has frozen a natural process and unfolded a map of a single moment in its development. It reflects the evolution of woodland itself. One suite of trees can be settled in a wood for centuries until soil and weather condition no longer suit them and they're succeeded – if humans permit – by other species. What happens when the dynamic phase of this cycle is shut down indefinitely?

*

Given its inherent invisibility – buried deep in the interior of trunks – I never thought I would see spalting 'in the wild'. But in June 2014, in a wood on Berkhamsted Common, a huge beech pollard that I'd known since I was a child was blown down, and on part of the newly prostrate trunk the bark had been ripped off, revealing a patch of bare wood covered with that characteristic mottle of hieroglyphs and enclosures. I'd paid my respects to the fallen veteran just a few days after its collapse, and for a while wasn't sure if I'd found the right tree. The space it occupied had changed dramatically, not just because the beech was now cleaved down the middle and lying like two wooden cliffs on the forest floor, but because of the light, reflected off gashed wood and tree surfaces that hadn't seen the sun for a hundred years. Much of the arcane culture of the upper reaches of the tree had fallen into plain sight – birds' nests, the healed edges of ancient wounds, scores of old graffiti. One read *18.V.44*. The characters were incised deeply enough to be Victorian, but I think this was 1944, when homesick US airmen were stationed nearby, and carved their memorials on the receptive bark of local beeches. And now I could get close it was clear that though the *Kretzschmaria* fungus was responsible for the spalting it wasn't the chief cause of the tree's demise. There were brackets of *Fomes fomentarius* along the trunk, as big as carthorse hooves. This is one of the most debilitating parasites of beech, turning the heartwood into something resembling crumbly cheese. In what is one of the least touched beechwoods in Europe, the *réserve biologique* at Fontainebleau, its brackets mount the immensely tall and uncut trunks and crash to the ground like meteors as the trees decay. There is no management in this wood save the exclusion of the public.

The wood where the pollard had collapsed has a different history, which echoes, on a landscape scale, the dynamic narrative inside the spalted branches. There was an extensive open beechwood, a wood pasture, at Berkhamsted long before the Norman invasion. The local commoners put out pigs and lopped the trees for firewood. The invaders appropriated the entire area, put in their own pigs (Domesday listed it as 'wood for a thou-sand hogs') and in the mid fourteenth century enclosed a large section as a hunting park, to which they introduced a herd of fallow deer. Over the centuries that followed the wood, called 'Del Frith', shrank as patches were illegally enclosed and trees looted, first by the Crown and then the Duchy of Cornwall. But it continued to be an assemblage of beech pollards in grazed heathland, a rich cultural landscape but biologically in a state of arrested development.

In 1866 the land-grabs reached a climax, when the landlord, the young Lord Brownlow, illegally enclosed four hundred acres of the common, including Del Frith and its venerable pollards. The outcome was a historic moment in conservation. The recently formed Commons Preservation Society coordinated a brazen exercise in direct action which involved disman-tling the three miles of iron fencing. They also sought a court injunction to prevent any further enclosures on the common, and clarification of what rights existed over it. The case dragged on for four years, but in the end confirmed common grazing rights, and formed a decisive step on the way to granting the public rights of 'air and exercise' on commonland.

But one ancient right for which no evidence was offered was 'estovers', the taking of the 'lop and top' of the trees. Perhaps the practice was already obsolete as a result of cheap coal. Whatever the reason, by the last quarter of the nineteenth century pollarding had ceased in what was now called Frithsden Beeches. And a new destiny opened up for the wood. Without cutting, the beech branches grew thickly and elaborately, many progressing to the condition of secondary trunks. Large branches began to be shed, leaving rot-holes which sometimes became host to aerial tree seedlings. Fungi, beetles, wood-nesting birds proliferated in the zones of dead wood. In places the high outgrowing branches of adjacent trees fused, echoing the way the roots were linked by underground mycorrhizal fungi. By the last quarter of the twentieth century the uncut pollards became increasingly top-heavy, and there were casualties in most major gales. As new light flooded in, natural succession was unlocked for the first time in centuries. Tree seedlings began to colonise the gaps – dense thickets of beech in places, but also birch and oak and holly. Many were munched back by the descendants of the Normans' fallow deer, but a few always got away. By the late 1970s, Frithsden was an extraordinary, mercurial place. There were still large numbers of pollards, which were becoming increasingly Gothic. Flared root buttresses, cankerous paunches, snags, bosses and burrs abounded. The volume of fallen and standing dead wood (or 'fallow wood' as the Reverend Francis Kilvert usefully called it) was beginning to edge towards the 50 per cent that is the norm in the near-natural high forests of Eastern Europe. All kinds of natural cycles began to re-emerge – not just the replacement of pure beechwood by a tree population more varied in age and species, but the constant metamorphosis of live wood into humus. Over the course of a decade or so I was lucky to witness one particular tree go through this entire cycle. I called it the 'Praying Beech' from the way two branch stubs had fused like a pair of clasped hands. One summer it was split open by a lightning strike. Bees nested in the hollow gash. A while later it was toppled in a storm. For the next few years it was host to a legion of wood-decaying fungi, a kaleidoscope of coral spots, dead-men's fingers, porcelain tufts. Then, one autumn, I was caught in a ferocious rainstorm nearby, and watched the trunk start to liquefy in front of my eyes. I wrote down what I saw, conscious of being a privileged spectator at something wondrous: 'The rain was hammering drills of water at the already rotting trunk. Flakes of bark fell off, then threads of fungi and sooty dregs from the lighting-charred heartwood, and essence of beech dripped onto the woodland floor like the oil from an alchemist's still.'

It hardly needs saying that this theatre of transformation with its craggy wooden cast was not to everyone's taste. I once tagged along behind a party of foresters and landowners on a tour of Frithsden Beeches. They were outraged that what they called a collection of 'mutilated freaks' had been given living space for so long. The consensus was that they should be cleared away and replaced with proper, productive trees. This became a serious suggestion to the current owners, the National Trust, accompanied by dark hints about the dangers to the public of derelict trees. The Trust, well aware of the local public's affection for the beeches, warts and all, put up notices instead: 'These very old pollarded trees and associated deadwood in this area are being managed for their nature conservation and historic interest. They are liable to shed branches and the public is advised to keep to the waymarked rights of way.' In fact, beyond a little clearing of fallen branches from pathways, these notices were the limit of the Trust's 'management'. After the catastrophic mismanagement of woods across England following the great storm of October 1987, non-intervention became Trust policy on wooded commons.

The focal point of the whole wood was a tree known as the Queen Beech. It was an immense pollard, maybe four hundred years old, with low-slung main branches that skirted the ground and gave the whole organism the air of a giant squid. It was the matriarch of the wood, and had such a low centre of gravity that I always imagined it indomitable. But this was the tree that split open in June 2014. And looking at the panorama of its levelling it seems to me a still life analogous to my spalted bowl, a snapshot of a moment in woodland evolution, evocative of past history and suggestive of possible futures. I find it hard to feel much sorrow for the fallen Queen, so *present* does it still seem to be. The surrounding trees show hollows in their crowns where they had been shaded out, which will become the templates for new growth. The great wedges of fallen wood, the spalted trunk, the graffiti, will endure for decades. Even the Trust seems to view the change positively. It has put up another notice a short distance from the wreckage saying: 'This famous tree has entered the next stage of its existence' – a sentiment that would have been unthinkable on public display even a decade previously. In the shadows I can make out burgeoning clumps of young birch, beech and holly around the dark hulks of earlier tree falls. They show the irrelevance of deliberate planting here, and the likely future of this sunlit space. In front of them the Queen's roots are hard and rippled, and remind me of stromatolites, the compacted clumps of silt and cyanobacteria that were some of the first living communities to form on dry land more than 3 billion years ago. Below ground their mycorrhizal fungal partners will already be linked with the new trees, trading nutrients and information about soil conditions in what has been called 'the wood wide web'.

Just for a moment everything seems perfectly poised for the next step. What should we do?

PART FOUR

The New Nature Writing

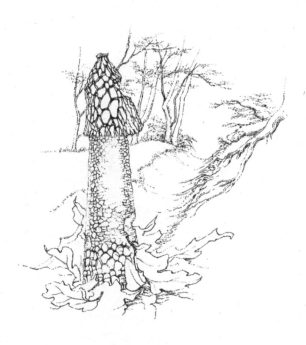

In the autumn of 2007 media pundits spotted what they thought was an emerging trend in literary non-fiction. They dubbed it 'the new nature writing', implying that it differed from some 'old', more purely descriptive nature writing, perhaps what one might call 'natural history writing'. In fact the trend consisted of just six titles. Mark Cocker's *Crow Country*, Robert Macfarlane's *The Wild Places*, Roger Deakin's posthumous *Wildwood* and Jay Griffiths' *Wild*, had been published earlier that year. Kathleen Jamie's collection of essays *Sightlines* had come out in 2005, at the same time as my own *Nature Cure*. They were a disparate bunch, in both style and content, but did have some important things in common. Each of the texts was presented as a first-person narrative, and didn't exclude the writer's interior reflections and subjective feelings about their notional subjects. Each was sensitive to the character of place and landscape, and the way they shaped the imagination. They granted subjectivity to the natural world too, exploring the extent to which wild organisms, rather than being passive victims of blind instinct and human enclosure, had agency, and created their own life stories. Science, for the most part, wasn't seen as the enemy of all this subjectivity but as a tool for validating it.

Just why all this was thought to be new is puzzling. There is a long tradition of previous British writing of this kind, stretching back through writers such as Kenneth Allsop, J. A. Baker, Edward Thomas, Henry Williamson, W. H. Hudson, Richard Jefferies, through the Romantics right back to Gilbert

White of Selborne. The pundits, enquiring what might lie behind this apparent new surge in titles, noted that four of us lived in East Anglia, and wondered if it was a reaction to the industrial farmlands of 'the Bread Basket of England'. A more humdrum consideration was the fact that the four of us knew each other as friends, and shared our ideas. The books were a kind of conversation between us, about the nature of landscape, ecological politics, the role of the writer. But if there was an overriding cause it was the arrival of a historic moment, when a generation brought up on American nature writing, and the intense but unselfconsciously personal work of authors such as Annie Dillard, Barry Lopez, Gretel Ehrlich and David Quammen, finally caught up with the UK's growing ecological crisis.

From 2007, the British school proliferated, and attempts to group them into a single genre began to look strained. There were explorations of landscapes, seasons and individual creatures. Bees, otters, butterflies, orchids, even moles ('Catch one and find yourself in nature') became the subject of quests, with publishers encouraging authors to frame their texts as 'journeys' of personal discovery. Increasingly the focus of the books shifted, so that explorations, and healings, of the self began to overshadow accounts of the nature that had prompted them. By the end of 2015, 'new nature writing' was beginning to look like a distracting and redundant description, as the output morphed into a subcategory of memoir writing with a rather unusual cast of characters.

That summer the *Guardian* writer Steven Poole had written an opinion piece denouncing the genre, not for its increasing introversion but, oddly, for being a new manifestation of the pastoral tradition, in which middle-class metropolitan authors indulged their fantasies about a vanished Arcadian landscape. This seemed a bewildering travesty to those of us in the business. I wasn't one of the writers named as guilty of various infelicities, so after some discussion with those who were, we agreed that I was best placed to pen a reply. It follows, together with reviews of a couple of the best books of the first wave of the 'new writing'.

Fantasists of the Fields?

First published in the Guardian, *2015*

Not since John Clare lambasted Keats for metropolitan sentimentality has there been such an attack on the integrity of nature writers. In a letter written in 1830, nine years after Keats's death, the poet of the fields accused the cockney Romantic of portraying 'nature as she ... appeared in his fancys & not as he would have described her if he had witnessed the things he describes'.

Steven Poole's sneer at modern nature writing echoes Clare's jibe, though without his authority. What Poole sees, lumped together under that portmanteau tag 'new nature writing', is at root just urban whimsy, pastoral nostalgia, 'a solidly bourgeois form of escapism'. He singles out half a dozen writers for particular derision, but his piece is plainly aimed at all of us who have tried to explore the complex relationships between humans and the natural world. Nature? For Poole it's just inanimate stuff, mostly malign. Not a place for civilised humans or serious existential exploration. Searchers in the forest are fanciful elitists, possibly proto-reactionaries.

Perhaps we have brought this criticism on ourselves. All that pious talk about 'reconnecting with nature', when we are all connected with it in every breath we take and every gene we inherit. All those books (I have written one myself) which see in the natural world some respite from emotional damage. There is a simplistic, self-indulgent tendency here, and there are legitimate questions about the way that nature has been commodified in such writing as day-trip redemption, green Prozac. But that wasn't the core of Poole's argument, which was that folk such as us – middle-class intellectuals for the most part – were disqualified from writing about nature, which was properly the prerogative of those who owned and worked the land, and therefore understood 'the way we do things here'. This was a resurfacing of the obsolete notion that, in Britain at least, 'nature' and 'the countryside' were one and the same, the proprietors of the latter being the rightful colonial masters of the former.

I've worked as a nature writer for forty years, and though I wasn't personally incriminated, I felt traduced by it, as were many of my friends and fellow writers. I can only speak for myself, but I know enough of the way this area of literary exploration has been evolving to raise a voice of complaint on behalf of my colleagues. To begin with, the genre is largely a recent invention of the media. Can one really lump together Caspar Henderson's precise yet endlessly astonished modern bestiary *The Book of Barely Imagined Beings* and Tim Dee's ornithological memoir, *The Running Sky*, with its audacious imagining of the polyphonic roost flights of starlings as the score of Thomas Tallis's motet *Spem in Alium*? Current nature writing is the broadest of secular churches. Oliver Morton's engaging personal saunter through the world of photosynthesis, *Eating the Sun*, might be more properly labelled imaginative science writing, just as Robert Macfarlane's literal wanderings in *The Old Ways*, in which footfall and thought-fall echo each other, is a variety of imaginative travel literature.

Yet there they are – we all are – grouped as a kind of cult, the eremitic 'new nature writers'. Most of us prefer to think of ourselves just as writers, who simply wish to embrace a rather larger than usual cast of characters – the other beings and landscapes with which we share the planet; and to respect them as *subjects* in narratives, not simply as objects. But nature as active subject implies relationships, not only between living organisms but between them and us, and the difficult work of balancing truth to one's emotional and imaginative responses with truth to an organism's own life.

All this, and the assumption that there is something close to 'mind' in nature is anathema to Poole. Might he accept 'intelligence' instead? Today, a neighbour brought me a goldfinch's nest, whose foundation was a perfectly selected and fashioned circle of chicken wire. It wasn't strong enough to hold the nest up in gale but it was a great idea. And all this last hot week the oaks have been putting out their pinkish lammas shoots, responses to foliage eaten by predatory insects. If there is more predation the oaks will begin talking to each other through airborne pheromones, which increase the bitter tannins in their leaves and hopefully deter further attacks. Intelligence in nature is this anciently evolved capacity to adapt, to be creatively resilient.

Attempts to move from human-centredness to more inclusive and empathetic points of view are philosophically and practically fraught, involving the thorny question of whether language (and therefore our writing) is complicit in humans' supposed alienation from nature. My own view is that language, one of our special gifts as a species, is part of our ecological connection to the world. Jonathan Bate, in the UK's first indigenous work of eco-criticism, *The Song of the Earth* (published five years before the first tide of new nature writing in 2005–7), wrote that 'the dream of deep ecology will never be realised on earth, but our survival as a species may be dependent on our capacity to dream it in the work of our imagination'. Which makes Poole's suggestion that modern nature writing is nothing more than retreatist pastoral, intended 'to whitewash the non-human world as a place of eternal sun-dappled peace and harmony', even more baffling. Has he not read Kathleen Jamie's sombrely beautiful and unflinching pathology-lab meditation in *Sightlines* on mortality, and on disease as an intrinsic part of nature?

To the extent that nature writing has a common spring, it is defiantly anti-pastoral. It emerged not out of a desire to return to some ruralist golden age, but to repudiate such fantasies: the tweeness of 'country lifestyle' magazines, the technically obsessed imperiousness of natural history television, the belief that agriculture – and its colonial embodiment, 'the countryside' – are unimpeachable sources of moral value. Hence the passion for the unfarmed wild, for the small, the particular and the local, and for the affirmation that 'new' nature writing is not new at all, but embedded in a long tradition. And though mindful walking is currently the fashionable route to revelations about the self and the land, for much of the past two centuries it was rootedness, staying put, which was regarded as the key. The Reverend Gilbert White, whose *Natural History of Selborne* (1789) was the first work of literary ecology, spent almost his whole life in that small Hampshire village. His one book was, for its time, a revolutionary exercise in modernism: non-fiction organised in the epistolary style of experimental eighteenth-century fiction, and daring to pay the same attention and respect to crickets as to churchgoers. John Clare, whose anthems of solidarity to his fellow commoners of all species are another beacon, panicked when he was away from Helpston and 'out of his knowledge'. Ronald Blythe (no mean nature writer himself) has lauded Clare's 'indigenous eye' and 'his extraordinary ability to see furthest when the view was strictly limited'.

The fact that there is now a flourishing strand of urban nature writing (e.g. Paul Farley and Michael Symmons Roberts's *Edgelands*, and Rob Cowen's *Common Ground*) is proof of the absurdity of Poole's pastoral charge. As I write, in a blessedly hot July, that icon of the urban wild, the swift, has returned to its brick-and-concrete parishes. Where swifts disappeared to in the bleak weather this May and early June is as tantalising a mystery as their return is a benediction. ('They're back,' Ted Hughes wrote, 'which means the globe's still working ...') Now the dark scimitars will be scything across north London back-to-backs, screaming at knee-level through the alleys of Granada, braving urban war zones across the Middle East. I once saw their flickering shapes crossing the commentator on a live TV broadcast of the shelling of Beirut. 'Choose life,' they screamed. Finding words to bridge that divide between the otherness of nature – a swift sleeping on the wing five thousand feet up – and the immediacy of the rush of wings past your face is what most nature writers are striving to do, not wallow in some vanished rural Arcadia.

Patrick Barkham on Butterflies

First published in the Guardian, *2010*

Vladimir Nabokov once wrote that he loved to 'drop in, as it were, on a familiar butterfly in his particular habitat, in order to see if he has emerged, and if so, how he is doing'. This was an untypically chirpy remark from the doyen of literary lepidopterists, and not quite in the mood of *Guardian* writer Patrick Barkham when he set out on his own odyssey. In 2009, strung out by urban angst, he decided to try and see every one of Britain's fifty-nine species of native butterfly in a single summer. As an achievable task, slugs, at twenty-three species, would have been easier, but might not have had such an appeal. As it is, *The Butterfly Isles: A Summer in Search of our Emperors and Admirals* reads as a compelling example of the 'quest' mode of nature writing.

Why do men – especially men – take on such acquisitive challenges? 'Because they're there' would be an odd answer for such beautiful, myth-bound creatures as butterflies, and smack of bathos. But Barkham has a more powerful motive. For him the universe of butterflies is like Alain-Fournier's *Lost Domain*, a place of elusive and endangered beauty, charged with the lost freedoms and magic of childhood. In a brilliant opening chapter, he recalls a summer's day when he was eight years old on the north Norfolk coast. He is butterfly-hunting with his dad (a heroic character throughout the story) when they spot 'a grey arrow flying low over the clumps of marram'. It resolves into a Brown Argus – tiny, unprepossessing, difficult to distinguish. But when it spreads it wings it is a revelation: 'A deep chocolate colour spread from the orange studs bordering the wings right into the soft brown hairs of its delicate body' – a change as astonishing as the metamorphosis from caterpillar to butterfly.

Nearly thirty years on, he is still in thrall to the paradoxical beauty of insignificant things. It's a magnetism that depends partly on particularity (which makes his desire to notch a full suit perfectly understandable) but also on 'fit'. His evocative descriptions of Norfolk's baking sand dunes, and the Argus's flight through them, are intimations of a deeper vein of connectivity that unfolds during his summer's pursuit.

Landscapes are powerful presences throughout the book. He begins in the scrubby wasteland outside Bullingdon Prison in late February, searching for the minute eggs of the Brown Hairstreak. 'I lightly held a slender branch of blackthorn, turned it gently to the light and systematically ran my eye along it, checking at every joint of a twig.' The Hairstreak's emerged imago is the very last of the fifty-nine species he sees in August. In the months between he criss-crosses Britain, recording his search in this same vivid, unapologetic voice, wonderfully catching the jizz of these ethereal creatures. He hopes, like Tove Jansson's Moomintroll, to see a lucky yellow butterfly, a Brimstone, as his first of the spring, but is baptised instead in his mum's garden by a Small Tortoiseshell, 'the Labrador of the butterfly world'. In May he is caught up in the epic migration of Painted Ladies, an extraordinary diaspora that takes them from North Africa to Iceland. On Start Point in Devon he and his dad find Pearl-bordered Fritillaries, distinguished by a single lustrous silver cell on the hindwing.

The minutely detailed patternings of butterflies are one of their mysteries, as the sexes find each other chiefly by smell. The male Adonis Blue becomes so intoxicated by the female's pheromones that he punctures the pupa and mates with the female *before* she has hatched. Humans can sometimes detect butterfly odours, too. Barkham is tickled by the image of venerable professors 'inhaling deeply over [a] prone downy body', and reporting that the Meadow Brown smells like an 'old cigar-box', the Wall Brown 'heavy and sweet, like chocolate cream'. And some butterflies can indisputably smell the odours of the human world. Picture this scene in an ancient hunting forest in the Midlands. Five tables covered with white cloths, and set with paper plates of rotten banana, dead fish, and Big Cock shrimp paste. The maître d' is the National Trust's rumbustious butterfly adviser, Matthew Oates, sporting a purple notepad and a purple band on his sun hat. The feast is an art event run by a local gallery, and designed to attract the fabled Purple Emperor, whose Gothic glamour is enhanced by the fact that it is, so to speak, more of a necrophile than a nectarphile.

Oates is one of many extravagant butterfly lovers (including Grimaldi and Churchill) to flutter through the book, but is most important because he introduces Barkham to the Welsh concept of *cynefin* – 'a place of personal belonging'. Butterflies have intense *cynefin*, which is why they are so vulnerable to change. The larvae of the Large Blue, whose life cycle and rescue from extinction makes for the most extraordinary story in the book, can only exist in intimate partnership with one particular species of red ant.

And in the end Barkham discovers his own sense of *cynefin*. Suffering from 'butterfly burnout' and dumped by his girlfriend, a reluctant 'butterfly widow', his senses – as they can in times of distress – start bristling like antennae. He finds he has become a dowser. Lost in a dark and rain-sodden Surrey wood, inexplicably scared, he discovers his grail, the ghostly Wood White, 'the shape of a droplet of water, hung on leaf nearby'. He understands that its 'ability just to be, and live fully in the present' – and in the present place – is exactly what he gets from butterfly-watching. He has become butterfly-brained – not, in his case, a term of derision, but a metaphor which exactly catches the way his darting and discursive narrative reflects the lives of his familiars and his sense of empathy with them.

Jay Griffiths on The Wild

First published in the Guardian, *2007*

One night in the Amazon basin, out of her head on the local hallucinogen *ayahuasca*, Jay Griffiths shape-shifts, and enters the mind of a jaguar. She finds herself prowling the streets in Oxford, and roaring in 'untetherable rage' at the Bodleian Library 'which houses with such care all the dry knowledge of years, *while the Amazon burns*' – and which now, of course, houses her own incandescent book, itself set out so carefully in a hundred thematic sections. This is just one of the paradoxes and contradictions that haunt this kalaedoscopic narrative, which is nothing less than a seven-year odyssey after that fugitive quality 'wildness', the anarchic vitality that drives natural creation, and, as Griffiths sees it, is the wellspring of the human imagination. *Wild* is not just about the knowledge of nature, but the nature of knowing.

But isn't wildness obsolete, irrelevant? Isn't it what civilisation was meant to defeat? Civilisation almost has defeated it, but Griffiths's impassioned argument is the same as Henry Thoreau's: 'in wildness is the preservation of the world', and that we lose touch with this protean quality in minds and beings at our peril. She could have followed her quest by sitting quietly in her garden for seven years, minutely contemplating the lives of its indigenous insects. Instead she follows a well-trodden path into some of the world's great wildernesses of forest, ice and rock. This isn't out of bravura, though she is a brave woman. Nor from that sense of pious masochism that drove Christian missionaries (and some modern-day adventurers) into these territories, and who are most satisfyingly speared, boiled and damned throughout her book. She chooses these places because she cannot agree with the conventional wisdom that humans are a stain on the wilderness. She sees us – or at least our feral souls – as part of nature, linked by our shared history and genes. But despite its proclaiming of this common ancestry, *Wild* is not an inclusive book, and does not much consider the natural connections of modern humans, which are presumed broken. Griffiths is most interested in what are now called 'indigenous peoples', who have lived in their niches for millennia. She wants to read the constitutions of their 'ecocracies'.

In the Amazon she goes mind-trawling with shamans, and introduces one of her central themes: the wild energy of language, which she sees as a direct mental correlate of the luxuriance of the natural world. The Desana people have words for intermediate colours in different kinds of light. But then the English have fifty different vernacular names for the foxglove: the special intimacy awarded by 'indigenousness' is not as absolute as she sometimes argues. Her excursion into the history of the *Oxford English Dictionary*, which was conceived as a project to tame the language but ended up as a riotously proliferating organism – 'glorious proof that language was unfixable and nomadic, wildly profuse and forever free' – is typical of her discursiveness. Her pages are like tree roots (etymologically connected to *radical* and *route*), wandering from mound to hollow, absorbing the nourishment of often very unwild academic disciplines – history, anthropology, geology – and sprouting into exhilarating prose.

She describes a whale hunt with the Inuit with a mixture of intense detail and deep-rooted Western nausea, and witnesses a people poised on a paradox:

following an age-old survival practice and despising 'the people who change nature', while buzzing about the ice with automatic rifles in American SUVs. In the Arrernte Desert in Australia she walks songlines with her Aboriginal companions, across dunes the height of a three-storey house. She sees mirages and the intricate network of canny, sand-wise organisms – the singers of the songlines – with the intelligence of a naturalist and the originality of a visitor from another planet.

There are many kinds of traveller in this book. The story is stalked by religious evangelists and mining corporations and military agencies who see nature not as our origin and the repository of the world's living energy, but as a state of sin or a bloody nuisance. Diving with the Bajo 'sea gypsies' off Indonesia, she recalls the US Navy using dolphins, with bombs strapped to their backs, as proxy commandos. She fantasises that whales may be engaged in song cycles thousands of years old. And in what is, for me, the most beautiful passage in the book, she rhapsodises about reef fish as if they had been alchemised out of music and phosphorescence. Yet, against her view elsewhere that names are a kind of entrapment, she dubs them all, recognises their individuality. Water is her natural element. She relishes its flux and femaleness, its refusal to be contained, its Promethean souplness, its joining of primordial sea and human body. Her tongue-insomewhere field guide to her own genitals makes them sound like a World Heritage wetland.

Wild is like that – sensuous, cocky, magnificent, overwrought, maddeningly contradictory. Sometimes I wished Griffiths *had* written of her own garden. Her argument that the wild is our true home, its rhythms and archetypes embedded deep in our cultural and genetic inheritance, is weakened by the unattainability of the distant 'self-willed lands' and ways of life she chooses as her arena. It is as if the wild is a privilege, or a dream. But of course it is everywhere, in the scream of swifts over cities and the worm under the plough. ('Beneath the pavement, the beach' was the slogan in Paris during the 1968 uprisings). Everyone is an indigenous person in the place where they live, however shallow or discontinuous their roots.

Griffiths implicitly underplays one other crucial point in losing herself in the anarchic energy of the wild. Unlimited wildness, free-form chaos, leads only to dissolution. Wildness survives by temporary settlement into pattern and order. The solitary mutation becomes the self-perpetuating species. The shaman's trance is managed by the dogmas of ritual. The wild howl becomes the written sentence.

I sense that Griffiths understands this, and simply feels that the ordered part of this cycle has been in charge for too long. And the central paradox of her book squares the circle. The puffs on the cover make much of its 'wild' style. But in fact its prose is impressively disciplined. It's often explosive and exciting, but in its Joycean wordplay, its scholarship, its ironic wit, its crafted cadences, it is stylistically a perfect example of a sharp and organised Western intelligence. As Gary Snyder advised in 'A New Nature Poetics', writing should 'be crafty and get the work *done*'.

We can't go 'back to nature'. But our best hope might be to go forward with it, using the richness of our imaginations and knowledges. *Wild* is an important contribution, an oratorio to animism thankfully transcribed in a very readable notation.

Surface Tensions

First published as an essay on spring in the New Statesman, *2014*

One of the associate tribes of nature writers, the psychogeographers, have been rebranded less dizzily as 'deep topographers' – or so the cognoscenti tell us. The BBC's arts correspondent Will Gompertz is making a film about one of their aspiring new acolytes, and is asking me if I see myself as one of their company. We're sitting on a bench in the Oxford Botanic Garden, surrounded by flowering irises and service trees, and I answer, too snappily, no, I'm a *shallow* topographer. It's a smart-arsed, irritable reflex at these tiresome abstractions, but I realise that I'm serious. I try to explain how, for me, landscapes are paramountly about their present life, their vivacious, protean, membranous surfaces, not some intangible, semi-mystical undertow. By lucky chance there's a visual aid on tap. From where we're sitting, the gate of the Botanic Gardens, built in 1633, was intended to perfectly frame the Great Tower of Magdalen College, and form a kind of Age of Enlightenment ley line. The wild card intervened, and the local builders misaligned it by ten degrees.

I'm being disingenuous, of course. Landscapes and nature work by a constant jiggle between deep pattern and capricious process. Solid rock meets persistent weather. Ephemeral greenleaf generates hardwood trunk.

Instinct negotiates with opportunism. Above ancient seasonal rhythms and inscrutable connectivities, life skits about like a cursor on a Ouija board, guided by chance and exuberant inventiveness as much as deep-rooted imperatives. And especially so in spring. Gretel Ehrlich, gazing over the Wyoming Hills at falls of migrating finches, falls of hail, crashings of orchard branches, concluded that in spring 'the general law of increasing disorder is on the take'.

I get disorderly and fidgety too after the months of rutted inertia, and wait for that day in early March when there is a kind of pre-spring overture, when the light seems to open out, lose the brittle clarity of winter sunshine and dust the leafless landscape with the merest hint of pollen. It happened on 2 March this year and, guessing where the action would be, I sped to the vast marshlands of north Norfolk. The atmosphere on the coast was electric. The sky was full of jitterbugging birds, wind-blown flurries of lapwing, chattering, cantankerous mobs of brent geese, flocks of golden plover, invisible until they turned in synchrony and the sun tinselled the undersides of their wings. I soon saw one reason for their restlessness. A juvenile peregrine falcon, driven by rapacious instincts, adolescent hormones and sheer devilment, was repeatedly scything at 150 mph through a shape-shifting plume of starlings – and missing every time. But I sensed another thrill running through the masses of birds. They were poised for their journey home, back to the northern tundra.

Do we still have this restless itch to move on somewhere deep in our own biology? We're touched by migration, bird migration especially, more than can be explained by the simple associations it has with the new seasons. The pioneering US nature writer Aldo Leopold envisioned the migration of geese as a kind of eco-poetic commerce, the corn of the Midwest combining with the light of the tundra to generate 'as net profit a wild poem dropped from the murky skies upon the muds of March'. Do the airy, insect-catching swoops of swallows and other migrants from Africa, so different from the movements of northern birds, sound faint cultural – maybe even genetic – echoes of that warm southern landscape from which the first nomadic humans emerged? Most of these annual visitors are in alarming decline, and we can have no idea of what we may lose if that link with our origins finally vanishes.

In the summer of 2010, just a few miles east of where I watched the peregrine, archaeologists discovered the oldest evidence yet of human occupation in Britain, a cache of flint tools probably 900,000 years old. They identified the likely makers as *Homo antecessor*, a group of hunter-gatherers who had risked the journey up from the continent to what was probably the northernmost habitable part of the European land mass. Happisburgh, where the find was made, is currently falling into the sea, but at that time was a bone-chilling boreal forest like northern Scandinavia. When the bitter 2010 winter struck, we locals took some pride in our antecessors' gutsiness.

I migrated to Norfolk myself in 2002, swapping beech-clad hills for windswept flatlands. With hindsight my journey seems as serendipitous as *H. antecessor*'s. It was driven by necessity but guided by chance – fortunate encounters, tangy memories of once-visited spots and longed-for creatures. Wafted north-east like a speck of spindrift, I ended up in the Waveney Valley, where I've lived ever since. I see it as home, but not as a place of new roots. It's not that I now feel rootless but that I seem to have become capable of briefly putting down new tendrils anywhere I go. As Bruce Chatwin argued, we're more nomadic as a species than it's politically convenient to admit.

But if I'm less deep topographer than landscape tart, I still have my manor, an entirely subjective parish that encompasses the land within a roughly ten-mile radius of my home. And every so often I beat the bounds, see what's up, what's about. I'm not, I hope, laying any kind of claim, just acting out that old warp and weave of nomadic curiosity and territorial affection. This year I had the fancy of making a perambulation on the first day of spring, I looked up the exact time of the equinox the night before – 20 March, 5.40 a.m – and just as one often does with a flight to catch, I woke exactly at that moment. It was barely light and the world looked flat and lifeless. I imagined the earth enjoying a brief moment of equipoise, just before it began to tilt again. What a hope!

I head west, out into the sand country. It's a mild, sunny day, but the drought is biting hard here. The ditches are empty and the hedges leafless – except that, thanks to another kind of migration, they're foaming with the white blossom of cherry plum, 'fools' blackthorn', brought here from the Middle East a thousand years ago. Much of the farmland looks as if it's been imported from a Martian agribusiness. Immense fields are shrouded in moisture-retaining plastic sheets, as shiny as mountain lakes. Pig farms have replaced cattle fields. Pigs in wooden pens, in corrugated-iron bungalows, in canvas marquees like a porcine Glastonbury. Nothing deeper in the topography here than a hog wallow.

I drive past the farm where in February an animal rights activist filmed the most horrific violence against stock the RSPCA has ever seen. A few days later the farmer, an honourable and much-respected man by all accounts, killed himself. There's been no identification of or even rumours about the workers responsible, but I notice that the ubiquitous billboards, urging us to support British pork because our animal-care standards are so much higher than 'abroad', have mostly vanished, replaced with 'Keep Out' notices.

A few miles on I climb over a fence, out towards a big sheep pasture, and hear the heart-stirring bubbling of curlews. I can't see them, but a buzzard glides overhead. They're now coming back to East Anglia, after generations of persecution by gamekeepers. Then I turn round and see a trapped magpie frantic in a cage. I want to spring it free, but the barbed wire around it is impenetrable. Impaled on the wire by their noses are a dozen shrivelling moles. William Blake, who abominated imprisoned creatures, might have seen this moment – the buzzard flying over the caged and sacrificial animals – as a kind of psychogeographical focal point, where two different kinds of values have collided.

This is edgy country, nervous of water shortages, EU regulations, and a public scrutiny unlike anything it has experienced before, and I'm relieved to move east and south into the clay country. It's a gentler, more intimate countryside, with small fields and smallholdings, old lanes and even older echoes. When I first came to live here I was browsing a large-scale map and was astonished to see that all the ancient features – green lanes, wood edges, field boundaries – were roughly aligned in a north-west/south-east direction. A few local historians had spotted it too. They calculated that this fragment of landscape, dating from the Iron Age, had a four-degree tilt to the west. It's invisible from ground level, so how it happened is a mystery. Thoreau had a theory that our species had a natural instinct to move in a westerly direction, following the course of the sun.

I follow my own instincts along this maze of lanes, through the village of Burston, where in the 1920s two London socialists defied the local gentry and clergy and set up a community school, supported by the villagers, which lasted till the outbreak of World War II. I find thin secluded valleys I'd never been in before, pass fuzzy commons, snail farms, otter streams, craft studios, a whole magpie ecology blessedly free of a cage. These valleys and wet patches have been the protectors of East Anglia's distinct sense of identity. They've kept the big roads away, and people come *to* East Anglia, not through it.

I end up in one of these miniature flood plain valleys, where I saw my first local barn owl, that ancient parish familiar. I haven't seen one here for two years, but just as the sun set one skewed out of a ditch. It flew off like the dismissive wave of a white cape, on an incomprehensible course over a dog-walking green. Barn owls do not fly high, but if it had, and had looked down on the parish I'd just circumnavigated, it would have seen a pattern that turned upside down my glib dismissal of deep topography. The surface membrane, inert, plastic, barbed and private; and flowing around and under it these thin meandering ribbons of life, first carved out at the end of the Ice Age.

I had moved to Norfolk in 2002, and the character of its subtle wet-ways was in sharp contrast to the open-faced downs and antic beechwoods I'd grown up with. They were cryptic, illusory, shape-shifting. It was obvious that the most intimate way to explore them was by boat. Polly and I had Coquelicot *(poppy in French) built for us in 2006, by the boatyard at Martham in the eastern Broads. It's an old-style Broads cruiser, but fully canopied and powered by a smart electric motor. We can nose our way along dykes to lookouts unreachable by land. Other creatures seem unfazed by the boat's quiet and slow-moving carapace. Great crested grebes swim deliberately towards us and surf the bow wave. Kingfishers have perched on the stern as we've breakfasted. Swallowtail butterflies fly over us as we lounge on the deck, a dither of primrose yellow and black lattice. One late-summer afternoon we had an osprey fish-prospecting directly above us, a spectacle reserved for the boat-borne. It dwarfed the harriers that rose up to mob it, and for the briefest of moments blocked out the sun. The osprey hovered, dipped, then went into a fifty-metre stoop into the water for a fish, pursued almost to the surface by the harriers. So much was happening that I abandoned my bins, and watched the drama unfold with my naked eyes.*

I relish the way we can use the boat to steal into other organisms' territory, but also the fact that it provides a kind of conceptual barrier to the world beyond. It is as if we're surrounded by a moat, or in a bathysphere. We can gaze out, but not, while we're on the water at least, get out. We can have the most intimate views of what lies beyond, but can't go crashing in pursuit. I find it salutary to have this constraint on our privileged viewpoint. A modicum of distance is not only respectful but a reminder of how different the lives and interior landscapes of our fellow beings are.

Coquelicot *gave me a new perspective on birds, and the cranes and harriers and barn owls of the Broads became fellow creatures in vivid and intimate ways.*

Turning the Boat for Home

First published as an essay on autumn in the New Statesman, *2015*

Mid-autumn, just before our boat goes into dry dock for the winter, has a special frisson on the Norfolk Broads. The reeds begin to bleach and reflect the sunsets, so that for a while the water appears to glow brighter as the dusk closes in. The last migrants leaving for Africa cross with the first arriving from the tundra, the swallow flying under the goose. This week the local kingfishers have reappeared, darting between moored-up cruisers and diving between their hulls. We've seen otters close to for the first time, one rolling right in front of the boat with a bream in its paws. But it's been a taxing year for many species. The wet and cold that have begun to typify Northern European springs have delayed and weakened migrant birds. The terrible and now well-documented collapse of insect populations has had a knock-on effect on birds' breeding fortunes. Swallows, cuckoos, willow warblers, nightingales are all crashing in numbers.

Does any of this matter? Is the fraying and fragmentation of species of any real consequence to us? The government seemed to think so when it set out its green agenda, and acknowledged that biodiversity was essential to the earth's survival and what it liked to call 'quality of life' (ours, that is).

Now DEFRA has asked the Law Commission to rationalise wildlife protection laws in the UK. The Commission's starting position isn't encouraging. The first duty of wildlife law, it has put on record, is to 'provide the framework within which wildlife can be controlled, so that it does not interfere with the conduct of human activity' – a principle which is equivalent to saying that the prime object of child protection law is to ensure the wretched infants don't get in the way of their parents' career opportunities. In an aside the Commission concedes that the law should protect individual animals from harm, but only if that harm is 'above a permitted level'.

It's not clear if these barbarous guidelines were dumped on the Commission by DEFRA. They certainly sit snugly with the government's social and economic project. But they may equally show the UK legal establishment returning to its default position on wildlife. The status of a wild organism in common law is as potential property. While it is free and alive, it belongs to everybody, or, more correctly, to nobody. But by being 'rendered into possession' – the legal euphemism for being killed or captured – it is turned into goods, the property of the owner of the land on which it's taken. The notion of wildlife as part of the family silver – private inheritance more than common heritage – melds seamlessly into the idea of it as disposable inconvenience, and many early protection laws carried an exception clause concerning 'interference with legitimate human activity'. But on this occasion the exception has been made the guiding principle.

As basis for legislation it is actively hostile to the conservation of our islands' wildlife, as well as offensive to anyone who regards living organisms as more than entries on a cost-benefit ledger. The problem is that we don't have an agreed alternative measure for the 'value of species'. Nor does that rather clunking term 'biodiversity' help much, with its implicit suggestion that species diversity always trumps abundance and connectivity. Our current attitude towards nature's 'usefulness' (the implicit converse of the Law Commission's 'interference with the conduct of human activity') is hardly appropriate. By useful we mean, of course, useful to *us*. We may have grudgingly admitted pollinating insects into the realms of the utilitarian, but not the predatory insects which attack the parasites of the pollinators. We allow agricultural fungicides to leach into the groundwater and collaterally damage 'useless' mycorrhizal fungi which feed local tree roots, and wonder why hedgerow oaks are withering ... The interdependence of species is too complex for us to make simplistic and anthropomorphic judgements about what is and what isn't 'useful'.

In September a huge fin whale beached at Shingle Street in Suffolk. It was thin and in distress and eventually died, despite Herculean efforts to get it back into the water. For a few days it became a kind of shrine, while the authorities worked out what to do with it. People flocked to the beach to see the sinuous carcass with its prodigious maw. They came out of a sense of wonder, or morbid curiosity, or simple melancholy. A great leviathan had lost its way and become embarrassingly dead meat. In the end utilitarianism triumphed. The whale was carted off on a low-loader to a processing plant, where its blubber was rendered down for biofuel. Were those of us who thought it would have been more fitting to bury the body on the shore guilty of sentimentality as well as serious impracticality? This not a 'conservation of biodiversity' issue: the loss of one fin whale is neither here nor there. But the fate of its remains nags us with another challenge: how we conserve the *meaning* of wildlife – which may underpin our so far feeble attempts to save it physically?

I'd like to argue that we should respect wild organisms for their own sake, because they're here. But I'm aware that this is a philosophical conceit, and that 'their own sake' is really code for 'my own sake' – or at least my aesthetic and moral satisfaction. The philosopher Edward L. McCord's book *The Value of Species* (2012) tries to find a compromise. He argues that 'individual species are of such intellectual moment – so interesting in their own right – that they rise above other values and merit enduring human embrace'. This raises utilitarianism to an intellectual level, but for me still fails to do justice to the richness of the experience of living in a world alongside other species.

Gliding home at last light on the Broads, the answer often seems self-evident. In October the pinkfeet return from Iceland. The great scrolls of geese unwind across the sky so high up that they make yet another plane of colour, their bellies lit pink by the sun long after it has sunk out of sight. But they're not remote in any other sense. The ebb and flow of their chatter, the calligraphy, the waving scribbles of birds ('taking a line for a fly', to misquote Paul Klee) speaks plainly about the company of one's own kind on great journeys. A few hours before they fly in to roost we round the corner in Somerton Dyke where the whirligigs begin. Everyone looks out for these engaging beetles, just a few millimetres long, as they drift about in flotillas close to the reeds. They shine in the sun like beads of mercury, and every few seconds the entire company bursts into a frenzy of high-speed, near-miss swirling, a waterborne roller derby. It's comic and touching and so far unexplained – except that, like the flights of geese, it feels intuitively comprehensible, a kind of dance about the companionship of crowds.

Whirligigs are ancient animals, whose family emerged more than 200 million years ago in the Triassic. They have no known predators, because of an extraordinary skin coating, which is a highly scented, toxic and anti-bacterial wetting agent. Their hind legs work like paddle-steamer wheels, and give whirligigs the highest acceleration of any aquatic animals. They do not 'interfere' with any human activity, neither are they in any way practically useful to us. And though they have undoubted 'intellectual moment' it's not at all clear why they touch us so. You round a corner, and there they are, at their usual address, and if they're not, you begin to worry, and miss them. This is nothing to do with anthropomorphism or manufactured empathy. It comes, for me, from something I can only describe as a sense of neighbourliness. Neighbourliness is not friendship. It's based on sharing a place, on the common experience of home and habitat and season. It might provide a bridge across that great conceptual divide between us and other species.

The Glee Instinct

Originally a Radio 3 Essay delivered before a live audience in Bristol in 2011 for a symposium on Earth Music.

I hadn't seen a wild crane before the mid-nineties. I didn't know of their secret breeding site in Norfolk (occupied since 1979) and had never been to any of their European redoubts. But they still cast a spell over me. I remembered seeing Mikhail Kalatozov's film *The Cranes are Flying* at an art-house cinema in Oxford, a touching love story set in Moscow at the close of World War II, in which the key moments are framed by chevrons of cranes migrating over the city, their echoing cries audible to everyone beneath. I'd read about the annual return of 50,000 birds to a lake in Sweden, an arrival that is announced on national radio and accompanied by a huge public celebration at the site. I knew that one-third of the entire European population spent their winters in southern Spain, fattening up on acorns, and thought this might be the place to see – and hear – them at last. So at the end of one January five of us, old birding friends, travelled down to the evergreen oak forests of Extremadura to meet them, and perhaps catch a first taste of the Mediterranean spring as a bonus.

Rain, the Spanish plain's proverbial element, was never far away, but the pulse of a new season was palpable. White storks were repairing their bulky nests on church roofs and electricity pylons. In the town of Trujillo their spring-cleaned twigs and guano began to coat the imperious statues of conquistadors. Out in the open forest, known as the *dehesa*, clumps of gnarled holm and cork oaks were putting on their first new leaves. Garrulous flocks of azure-winged magpies, with sky-blue wings and pinkish bodies, swooped between the trees. In the middle distance egrets and storks padded across the waterlogged grass, stabbing frogs with their dagger beaks. And scattered among them were groups of cranes foraging for acorns, mostly in family parties of three or four, red-crowned, bustle-tailed, decorously picking up their famously splayed toes, the *pieds de grus*. It seemed a harmonious panorama of the edge of spring.

At night the cranes roosted communally in lakes and reservoirs, and we longed to see – and hear – them fly out from their refuges. Early one morning we went to the Embalse de Navalcano, where some five thousand spent the night perched in the shallow water. It was still dark as we drove up across flooded grassland, past surprised barn owls perched on fences. The cranes were already stirring when we arrived, dancing and murmuring, eager to be off. We hunched down as inconspicuously as we could in the scrub. When they flew up it wasn't a sudden explosion, more a slowly mounting torrent, a scrabble of stick-birds barely visible in the dawn half-light. But they swept off to a tumultuous crescendo of sound. Cranes' calls are often described as trumpeting, but really there is nothing brassy about the sound. It is shrill, echoic, far-reaching. It has a touch of wolfish yelp about it, something almost percussive, as if the bird is striking a tense string in its throat. The clamour of five thousand birds calling as they flew overhead rippled across the sky like a peal of celestial bells. It was unmistakably a chorus, a conversation.

Later that day, our Spanish host took us to a forest hut for lunch. It was built of stone, with a single manna-ash trunk holding up the roof. There was an open fire at one side of the scatter of tables, and a bar with a corrugated-iron roof, from which hung slabs of dark Extremaduran ham.

As we were finishing our meal, a group of men near us began to sing. It was almost imperceptible at first, like the mutterings of bar-room gossip. Then the five men, Gypsy forest workers as it turned out, plunged into an electrifying outburst of flamenco. There were no guitars or clicking shoes, just those tense, plaintive, wavering lines of passion that seem to come from somewhere far behind the closed eyes. The men took it in turns to sing, each one trying to outdo the previous singer with more outrageously boastful lyrics, more elaborate curls and twists of vocal delivery. They would turn fierce single notes into poignant quarter-tone tremolos, utter gasps of ecstasy and anguish, sound Moorish and Irish all at once. Each boast and flourish was cheered and clapped by the others. There was a lot of hugging at the end of each turn. I suppose the whole recital was competitive in a playful way, but also a communal performance, a kind of musical relay. Then, unexpectedly, the singers began to enquire why *los ingleses* were not singing. Our consternation mounted as we realised the full paucity of our vocal skills compared to theirs. For some reason I cannot fathom now we decided on a rendition of 'Greensleeves' – perhaps because it was so perfectly English, and one of the few songs we all knew. So, led by the thin but plucky soprano voices of the two women in our party, we launched into this sixteenth-century courtly serenade. The flamenco team listened to our reedy performance with polite incomprehension, but applauded and bought us cider – one taste at least we shared. Then they started to offer us horses to buy, and we felt it prudent to say goodbye.

It was only later that I realised these two disparate musical experiences had things in common. One was that neither of them needed translation. Doubtless the cranes' chorale had encrypted information comprehensible only to them. Head west. Ignore ungainly bipeds crouching in the bushes. But its chief meaning was as clear as day. It was a celebration of a new morning, a communal shout of solidarity. The flamenco had content too, of course, if only we'd known a bit more Spanish. We took what we could from the body language: my sight-reading of women is incomparable; my arpeggios are like serpents. But what would have been understood by anyone on the planet (and maybe any crane) was its revelling in the energy and camaraderie and mischievousness of the voice.

One other realisation dawned on me, obvious enough in retrospect. Singing is infectious. Willingly or not, in whatever medium you fancy, you begin to join in. You nod your head, twitch muscles in time with the rhythms, form the shapes of sympathetic noises in the back of your throat. Singing occurs in all human cultures, and most of it is done in company. It suffuses the natural world, too. It may be semantically questionable to call all of it 'music', but group singing is an elixir of bonding, a dissolver of boundaries, a raiser of mood and generator of empathy, from the chatter of jackdaws going to roost and the chorus of humpback whales, to the crowd at Anfield and the Last Night of the Proms.

I'd learned something about the power of song when I was at my boys-only school. I was a timid teenager, sagging under the routines of an over-specialised science curriculum. Music was one of the few escapes, and got me into trouble for letting it eat into physics revision time. I played thrash guitar in the school skiffle group, and taught myself the rudiments of classical guitar, at least to the point of becoming a passable accompanist. I sang bass in the choir. The highlight of our musical year was a combined performance of Handel's *Messiah* with the local girls' school. We rehearsed on summer evenings in the parish church, the boys on one side of the chancel, the girls – unapproachable, but sublime in their green gingham dresses – on the other. We sang 'All we like sheep have gone astray' across the aisle at each other as the sun tipped down through the west window and the swift packs screamed around the spire.

All that adulation at a distance changed when I was in the sixth form. A young and enterprising maths master joined the staff and announced his intention of setting up an early-music choir, of both sexes. It was an audacious experiment, the first time pupils from the boys' and girls' schools had crossed the aisle and come physically together for an official activity. A dozen of us gathered twice a week in the late afternoon. We sang grand motets by Gibbons and Tallis and frolicsome madrigals by Dowland and Morley. In winter we turned to medieval carols and sang them in the lamplit porches of our teachers' and parents' houses. In the summer term after A-level exams had finished, when most of the choir were due to leave, we went to Oxford for the day, lashed two punts together and floated down the Isis singing Gibbons's madrigal 'The Silver Swan' ('who living had no note, when death approached unlocked her silent throat'). We would doubtless have wept if we'd been a little older.

The experience of singing in this choir transformed my last years at school, and was an emotional revelation. To form a *chord* with another voice, a girl's voice. To amble along a line of counterpoint, tongues tied. To begin to glimpse how personality displayed itself in sound and posture, and how the sultriness of one desirable contralto suddenly became unfrightening when we swapped vowels. I often left rehearsals red-faced and tingling – not just, I suspect, because those long Tudor phrasings made you hyperventilate.

Singing together is a physical business, with not just the vocal cords but the whole of the body involved. And it's in a mixture of sung sounds and linked physical gestures that anthropologists believe our social communication evolved, a synergistic stewpot of rhythmic noises, body signals, pitch-switching phrases they inelegantly call 'musilanguage'. You can glimpse this proto-music in children's play, and in their compulsion to chant and bob and mimic; and in adults' talk *to* children, that modulated, rhythmic, high-pitched cooing which is common to all cultures. From this primal lexicon, the theory goes, music evolved along one path, as principally an expression of emotion and social solidarity, and spoken language along another, as a more utilitarian channel for conveying information.

This division is also evident in bird vocalisations, which can be roughly divided into functional calls – alarms, contact notes, flight calls – and the more richly affective song itself, designed to affirm territory and attract a mate. But the division is porous. Many birds have simple songs – chirps, buzzes, grunts – which, to humans, sound no more elaborate than calls. The crane is an example. The mass chorus of cranes when they're flying together is, strictly speaking, a symphony of contact calls, not a song. But it plainly has a powerful emotional and social charge for the birds. I've heard the true song on the Broads, performed by a pair at the nest. It's a synchronised duet in which a two-note phrase is echoed by the two birds in such quick succession it sounds like the yodelling call of a single creature. (The interval, incidentally, is a minor third, which also occurs in all known human musical cultures.)

It might look as if birdsong in this sense is non-social. Male birds in the singing season are, precisely, attempting to make themselves individually conspicuous, proclaiming their territory and hoping to become sexually desirable by means of the elaborations and curlicues of their singing. Just what characteristics of a song make it appealing to a female bird isn't often questioned, and even scientists are apt to refer glibly to 'good' singers, as if the simplistic (and anthropomorphic) indices of, say, volume and tunefulness and versatility that impress us also impress female avian ears. But birds' hearing is so sensitive, and the amount of information compressed into a song so vast (in some species a hundred notes a second where our ears hear just one), that females will be responding to subtleties beyond our ken. Birds' singing organ, the syrinx, is capable of producing more than one note simultaneously, and when songs are slowed down, say four times, a much stranger and more elaborate sonic landscape appears, one which may be closer to that experienced by listening birds. It has something of the quality of mid-twentieth-century serial music, with bottom lines answering top, and frequent alternations between loud and soft. But that of course is just another anthropomorphic interpretation. The only scientific analysis I know of the relative success of, for example, different nightingale songs, suggests that those that sound the most elaborate to human ears are of those birds that have failed to find a mate.

But it's easy to forget that apparently solitary singers always have one listener – the singer himself. The ornithologist Charles Hartshorne argued as early as the 1930s that maintaining territory isn't a desperate, all-or-nothing affair like keeping possession of a meat bone. He suggests that there is time and room – and probably a need – for the singer to be sustained by the interest of singing itself, as a sensory and physical experience, and doubtless a generator of pleasure-producing chemicals. The biologist and essayist Lewis Thomas listened to a thrush in his backyard 'singing down his nose in meditative, liquid notes, over and over again ... I have the strongest impression that he does this for his own pleasure. Some of the time he seems to be practising, like a virtuoso in his apartment. He starts a run, reaches a midpoint in the second bar where there should be a set of complex harmonics, stops, and goes back to begin over, dissatisfied. Sometimes he changes his notation so conspicuously that he seems to be improvising sets of variations. It is a meditative, questioning kind of music, and I cannot believe that he is simply saying, "thrush here".'

One June afternoon in Provence I tried *listening* to a nightingale properly, not probing its song for musical or poetic resonances, but simply trying to hear what it was doing. It was very hot, and I'd taken a break from driving for a stroll through the fields. I hadn't expected a nightingale (it was late in their season) but one started up, hesitantly, in a thicket by the lane. I sat down in the shade, and just listened. It wasn't a loud singer, and certainly not a virtuoso. Its phrases were short, modest, and had a kind of inward, reflective quality. There were plenty of wevy-wits and grig-grigs, but no flaring crescendos of wew wew wew. But the bird persisted and the performance started to build into something very pleasingly structured and compact. I began to wonder what determined the order of the phrases. Was the bird consciously choosing each one, as if it were reading from a score?

But if not, was its brain firing them off automatically, much as a computer with a random selection program might do? Both alternatives seemed implausible.

Across the lane I could hear the shouts of children playing in the village. And the phrase 'absent-minded' drifted into my head – which of course doesn't at all mean the mind is absent, but describes a state where concentration drifts in and out of full, conscious awareness. As if to prove my point, the nightingale sang a quiet but elaborate phrase twice, as if it was suddenly pleased with what it had expressed. And I had an intuition (nothing more) that the bird was doing something analogous to me when I was whistling to myself, mixing ('absent-mindedly') mood and an innate skill with learned musical phrases, and then suddenly, momentarily, being 'pleased' with what I'd done. Maybe like most living things, a nightingale gives its most effective performance when its song touches and arouses its own pleasure centres.

Over the past thirty years I've spent a lot of time listening to and thinking about nightingales, wondering how it is that their songs seem like such *constructed* performances, and why, across the northern hemisphere, they have cast such a spell over listeners. Their song is not music, but it suggests musical tropes and structures. It's richly associative, and evocative of place and weather and seasonal moment. For me, it is 'as if' the bird is making expressive responses to these circumstances in a foreign language that we can nonetheless dimly grasp.

Three years ago I was out searching for nightingales in the Suffolk heathlands. It was the May full moon, but very un-springlike weather. At eleven o'clock the car thermometer registered 3°C. There was the faintest sparkle of ice on the heather, just as in George Meredith's bewitching nightingale poem 'The Night of Frost in May'. But there was not a single bird singing. I felt troubled that the catastrophic decline of nightingales –

some 90 per cent over the past fifty years – might have reached this ancient heartland. I walked to the end of the lane for one last try. At midnight, quite abruptly, one started up, in a dense clump of cottage-garden rhododendrons, a setting as incongruous as the temperature. It flung out a few icicle-sharp notes and then stopped. But it was a decisive statement. 'I'm here, *despite* ...' it sang, its terse, no-more-than-necessary statement seeming like a telegraphed bulletin on behalf of its silent – or absent – fellows.

But there is one reliable natural chorale. For a couple of months every spring we have the dawn chorus. Hundreds of different birds of dozens of different species, all within earshot of each other, begin the day with songs about their identity and mood and territorial loyalties. It might be tempting to call it symphonic in its diversity of sound and narrative movement, except that its different voices are completely uncoordinated and it has, or so we assume, no collective purpose. But I like to think it has something in common with that flamenco duel. Potential conflicts are resolved by the singing, and as in polyphony, each voice is respected as an individual.

An Owl for Winter

Originally a short article for the Mail on Sunday, 2010

The view from my study during these polar days (2010) has been like an illustration from a Victorian book of moral parables. The magnificence of the frost suggests the beauty inherent in all Creation, but there's an undercurrent of unease, a dark lurking, as if to remind us of the sin of complacency. Just yards from my window is a pond, dug four hundred years ago to provide the mud for building the house. For the first time its glassy surface looks thick enough to skate on – if you ignored the ominous air bubbles trapped in the ice. Behind it is a bird feeder we made from scrap metal in the form of a giant hogweed, where hunger is teaching birds like robins and dunnocks to cling to nut bags, something unthinkable a decade ago. Our trees have gone beyond the delicate lacework of hoar frost. They're like extravagant confections in spun sugar, and redwings and fieldfares, thrushes which optimistically migrate to us from Scandinavia in the hope of warmth and food, perch briefly on their tops like brittle ornaments. And beyond them, just out of sight, is a tumbledown farm building, where for three years a barn owl spent the winter. Its enchanting and strangely companionable driftings about the hedgebanks and fields I've come to regard as my home patch changed the way I cope with winter and how I view our puzzling relationship with non-human beings.

So perhaps I should add that this prospect is viewed from a warm room, with a lamp in the window – except that this isn't some romantic lantern, but an artificial daylight lamp, to help me manage the winter blues. I don't suffer from full-blown SAD (seasonal affective disorder) but I don't cope well with winter. It makes me anxious and claustrophobic, reminds me of parental illnesses, of a pre-central-heating childhood when water often had to be collected in buckets from communal standpipes. Dazzling frostwork, beneath the surface, is as deadly as the interior of a firework.

And the ambiguities of the view remind me, most painfully, of the great freeze-ups of the 1980s: the dead wrens lying in the porch, mistakenly believing that light meant warmth; the afternoon when a violent blizzard blew thousands of those hopeful redwings onto the Norfolk coast, and they lay dying in the snow, russet underwings slowly blackening as their last shreds of body-warmth melted the snow into tiny grave-spaces.

I moved permanently to Norfolk eight years ago, after that predisposition to winter gloominess turned for a while into clinical depression, and I needed to start my life anew. One thing I yearned for was to have barn owls as my companions again, as they had been when I was young. The memory of them beating past the poplar trees in the early spring – burnished golden wings against lime-green leaves in evening light – is one of few visual memories from childhood I can summon up with absolute clarity. I didn't see another in the Chilterns for thirty years. Their populations had been ravaged by agricultural chemicals and loss of hunting habitat. But in the 1990s there began to be sightings around Tring Park, an expanse of beech-wood and natural grassland about five miles from my home and less than one mile from my wood. The park had had an eventful and at times exotic history. In the late nineteenth century it was purchased by the Rothschilds, along with most of the surrounding countryside. Walter Rothschild, the errant son who preferred zoology to banking, stocked the park with emus and kangaroos, and rode about the district in a trotting cart pulled by zebras. In the mid 1990s it was bought by Whitbread, which had ambitions to turn the estate into a golf course and country club. Fortunately their scheme fell foul of planning regulations, and while squabbles about its future simmered (it was eventually bought by the Woodland Trust) the land itself regressed to an unkempt and seductively brooding waste.

I began to visit it in this fallow spell, relishing the sense of a landscape getting its breath back. The grass grew tall and silky, and prodigious plants began to appear. Then the barn owl rumours began. There had been sightings by the local allotments near the park, and one was spotted on a path, munching a vole. I took to going for dusk-time walks among the ragged hedges and overgrown paddocks that lay between my wood and the park. I never saw an owl on these meanders, but the summer smell of hay and the feel of rough grass on bare legs and the setting sun striping the beeches with their own shadows reminded me powerfully of evening walks as a boy. Our playground then had also been an abandoned park, and we had barn owls too, nesting in the old stables of the derelict mansion.

On a whim one July evening, I thought I'd do a vigil in the park itself. It was about 8.15, and I saw the owls as soon as I walked through the trees. They were quartering the bleached pasture, a pair of birds, one much darker than the other. I couldn't quite believe my luck – except that it wasn't entirely luck. I'd gone at the right time of day to the most likely hunting range, narrowing the odds. I leaned against a tree to watch them, the first close-to birds I had seen in the Chilterns since childhood. I was so thrilled I barely took in anything more than their luminous presence. They were larger and less white than I remembered, and not so weightlessly moth-like. They were flying low and choppily, and diving feet first for prey about once every thirty seconds. I didn't notice the intricate aerobatics of their hunting flights then, just the slight stutter and swell of the wings before they dropped. As far as I could tell each bird was catching something roughly one strike in ten, and after successful strikes they flew with their prey towards a long avenue of ancient beeches. I went and sat among the tall vegetation directly on their flight path, and soon they were beating past me only yards away, their pale faces and tremendous dark eyes echoing the luminescent flower discs of the hogweed. They seemed unbothered by my presence, and what was striking was the silence of their flight. Not the slightest thrum or hiss came from their broad wings, whose downiness had evolved precisely to dampen out their sound.

I went back to the park on several evenings that July, and found the hollow beech in which the owls were nesting. The tree had been split and scorched by what looked like a lightning strike, and the birds were gaining access through the fissure. They were never as open and demonstrative as on that first encounter. But once I glimpsed the young owlets. They had come to the entrance of the nest hole at very last light, and were hissing at their parents. The whole family appeared to be dancing.

That was the last time I saw them in the park. But they had sounded some resonance in me that I couldn't explain or elaborate on yet. It wasn't just their obvious grace and beauty, the way their pale wings seemed dusted with phosphorus in the midsummer moonlight, or shuttled through the low plates of mist after a hot day. It had something to do with the nature of dusk, and the paradoxical intensifying of detail that happens with the fading of the light. When there was less to see, what was there had an intense presentness. It was also something to do with the owls and me briefly sharing territory in a place we both knew intimately. It would be some years before I began to disentangle my feelings, but I knew that the birds had become important to me.

Barn owls' national decline continued through the 1990s. Poisoned by agricultural rodenticides, flattened by cars, the rough pastures where they hunted voles ploughed up, their UK population dropped from about 30,0000 pairs to no more than 5,000 by the beginning of this century. Even now, recovering as they are, they're no longer part of our everyday experience. The pale shawl floating across the hedge, caught in the car headlights; a startling glimpse of that heart-shaped face as the bird turns towards you on a post – they're now a local privilege. When I moved to Norfolk in 2002, I was taking up residence in one of their last British strongholds. The birds had hung on in the county's coastal marshland and low-lying pastures, and there were signs that the population might be increasing again. But it was a year before I saw one.

I was eventually tipped off by a friend, whose window cleaner reckoned he saw a white owl most evenings when he took the dog for a walk in a little valley at Botesdale, some three miles west of where I live. I knew the spot. Tucked under the village street, a tongue of rough grassland about a mile long edged both banks of a thin stream. There were patches of rush-studded swamp, solitary willows, half-hearted attempts at starting small plantations. I went there on a cloudless afternoon in early March, and settled down close to the stream by one of the willows. I sat there stock-still for an hour, becoming frozen and increasingly despondent. I watched a woodcock probing in the water, the stripes on its back wavering like the eddies on the surface of the stream. Then, just before dusk, the owl materialised, simply rose up from the grass in front of me, as soft and buoyant as thistledown. It was flying against the western sky, and the last light shone through its wings, marking out the dense primary feathers from the almost translucent wing edges. For a moment it seemed to have four wings, two in the day and two in the dusk.

I went back home with a head full of images like this. I was writing about my move to East Anglia at the time, and the encounter with the Botesdale owl went down immediately in extravagant phrases. 'It took off, its head like a quite separate creature riding shotgun out front.' 'It passed through the young saplings in the plantation, quickening its wingbeats to negotiate the gaps. It was winnowing the grass, threshing it for food.' I was rather chuffed by my poetic metaphors, and it was only when I first read this passage out in public that I realised its slovenliness. Any owl that bashed the grass like a thresher would rapidly condemn itself to starvation. Their hunting success and therefore their survival is predicated on stealth and silence. I'd masked the real bird behind a veil of superficial word-painting.

A few months later the owner of the house where I'd lodged when first moving to Norfolk told me she thought she had an owl roosting in her barn. I went to look and found the floor scattered with the pellets in which owls throw up the furry and bony debris of their prey. They looked as if they were coated in shellac, prepared for exhibition. I wished it had been around when I was living there. But some instinct nudged me, a sense that I knew exactly where it would hunt. The River Waveney runs close to the barn, through alderwoods and wet meadows and slaloms of scrubby dykes. That evening I went to the meadow nearest the road, the one most complicated with tree perches and bits of ditch, quite certain I would see the bird. It was perched on a hawthorn bush fifty yards away, looking directly towards me. When it flew out, against the dry reeds at a dyke edge, it showed off a back and wings striped in resplendent ginger. That moment, forgive me, I nick-named it Orlando, after the marmalade cat in Kathleen Hale's children's stories (and earlier than that a swashbuckling knight in an epic Italian romance).

I watched Orlando all that winter. He (the name I'd given it rather forced me into assuming it was a he) was an obliging bird, and a strict timekeeper. He kept to the same suite of meadows where I had first seen him, arrived there from the direction of the barn just after sunset and would fly close to me, as I stood in pilgrim reverence by the riverside. I began, dangerously, to believe I was becoming wise in the ways of owls. Then in mid-January something odd happened. I'd planned an afternoon stroll from our house on a clockwise route that would pass between arable fields surrounded by wide bank and ditch systems, and turn for home through a cluster of rough paddocks. I'd not gone more than a couple of hundred yards when I saw a pale shape billow from a derelict farm building. It was a *different* barn owl, much paler than Orlando, and for the next hour I had the eerie experience of following it, fifty paces behind, as it wafted along the exact route I had mapped out before leaving the house. I felt ridiculously empowered, as if I had not just predicted the owl's flight path but was conjuring it on its way. Still recklessly sentimental, I assumed its creamy pallor meant it was a female, and dubbed it Angelica, after the pagan princess that Orlando pursues in Italian romance.

For what remained of the winter I followed both owls obsessively, worrying about them in bad weather, and about what might have happened when they vanished for days on end. Most of all I wondered if they would get together, and one afternoon, I did a mad dash in my car to see how far apart they were and if I could glimpse them both in the time it took to do the drive. (I did and it was five minutes over a mile and a half. As I was parked up watching Angelica I found myself gazing into the cab of a similarly parked HGV, whose driver was also tracking her through binoculars. We exchanged knowing waves.) The two birds had different personalities. Orlando was a sentinel, fond of perching in trees and peering down for prey. Angelica was a stalker, a haunter of ditches and hedge bottoms. I once watched her fly down into a favourite meadow and lie (not perch) on the ground among a group of hares, all seeming to be comfortable with each other's presence.

Orlando met a miserable death that spring. I found his mangled but unmistakable ginger body by the side of the A1066, only a few hundred yards from the meadows where I'd watched him. I tidied his remains up as best I could, and what shocked me almost as much as his death was the smallness and insubstantiality of his body, an elfin form hidden in a magnificence of plumage. I could not drive past the site for weeks. When I next did, the roadside strimmers had been at work and all that was left were puffs of feather and splintered bone.

Later that spring I learned, from those who knew much more about barn owls than I did, that dark-plumaged owls are more often female, and pale birds male. I'd got their sexes the wrong way round, a deserved rebuke for my sentimental namings. But the bird I'd dubbed Angelica returned in the autumn to the tumbledown roost where I'd first seen it. I was watching one evening when it was driven into an oak tree by a group of tetchy crows. I crept up to the tree and leaned against the trunk. A few seconds later the owl flew slowly out across the field, but then veered back towards the tree, turned its head and stared straight at me before flying on. It seemed a gaze of curiosity, or annoyance maybe, about this creature that had been stalking its fields for nearly a year. But of course I hadn't a clue what it was feeling or even seeing. The huge, jet-black front-mounted orbs of barn owls mean that you can make eye contact with them, but your gaze dissolves in the dark void. It is impossible to say if they are looking at your eyes, or your feet, or that maybe even just a few yards away they can take in your whole frame. I can't say if this bird saw me as any different from other humans it had witnessed, but I was beginning to regard it as an individual. I'd written about barn owls as symbols of the dusk, talismans of 'good ground'. But I could no more reduce this intimate companion to the status of a symbol than I could regard it as an object. I wanted to get to know it. How did it cope in the rain? Why would it hunt the same beat at the same time for days on end, then suddenly choose to go round it the other way? Had it met other local owls, and might they get together? I was beginning to see the owls as my *neighbours*. It was a revelation, not just for the melancholic in me, but for the writer. For much of my working life I have been trying to find a way of talking about other organisms that neither reduces them to mechanical objects nor turns them into sentimentalised versions of ourselves. Neighbours are fellow creatures, but independent souls. There is no need for reciprocity. You share their territory – their parish – and often their fortunes, but you can care about them in full knowledge they may not even recognise you.

From then on I began to try to engineer encounters with barn owls across my home range. February and March are fruitful months as the female owls are feeding hard to get their bodyweights up to the breeding optimum (340 grams is the goal, according to the experts). I go out an hour or two before sunset. I love the dusk and find, maybe as a recovered depressive, something uplifting about the teeming life of imminent darkness. For a while I become an obsessive owl-dowser, poring over maps of the countryside round about for hints of the rough pastures and riverside grasslands that the birds haunt, and hoping that whatever I've intuited from their habits might guide me more precisely to where they're flying. Then I go there and wait. I get lucky about one trip in two, which feeds my conceit that I can suss these birds. But it's heartening when it happens, a token that in these particular places at least things are as they should be, grassland and vole and owl properly connected. I've found them in the grassy corners of elegant landscape parks and on overgrown commons. Once I saw one hunting *inside* an open barn. But I soon realised I wasn't being especially clever. These are the places where barn owls should be, and in an owl-rich region like East Anglia my success rate is what you would expect.

I once talked over my habit with my friend the radio producer Brett Westwood while we were recording a programme on owls, and he told me his own preferred watching style. The barn owls that touched him most were those chanced on by accident, glimpsed in the corners of fields, surprised on posts, wavering across remote lanes. These birds affirmed to him something about the owl's adventurousness and refusal to live according to our expectations. I felt suitably chastened. If I was honest with myself, I'd glimpsed more owls unexpectedly like this than I had on all my carefully planned and self-gratifying searches. Owls on improbable roadside perches: one on a Men at Work sign; another on a runner bean trellis at the edge of an allotment; a sunset-lit bird gazing from the top of a village signpost during the afternoon school run traffic jam. Serendipitous concurrences between two lifestyles. Many of these casual encounters happen at dusk or even in the dead of night. After years of looking you begin to develop pattern recognition for odd protrusions on wayside branches or a sense of fluttering whiteness behind a hedge. But it's never reliable. At a distance in the dark I've misidentified a tree-climbing white cat, a discarded trainer and most often the white gash of a freshly broken branch.

I send in records of all my sightings to the British Trust for Ornithology winter bird census, but confess that I don't watch like an inquisitive naturalist, hoping to learn more about their tastes in prey or breeding success. I've joined in ringing sessions at occupied nest boxes in summer, and watched the ladders raised to the box, the nestlings extracted and put gently into sacks and weighed, the metal rings clipped expertly to their spindly legs. And each time I am overwhelmed with anxiety, for what the parent birds must feel seeing their families disrupted like this, for the fragility of the young birds experiencing their first human touch. I have gingerly handled nestlings myself, terrified that I will hurt them and, as with Orlando's remains, been astonished at the smallness and lightness of the bodies inside their capacious downy fleeces. But this is sentimentality on my part. The birds are almost never hurt by the experienced ringers, and seem to forget their ordeal in minutes. What's learned is considerable. Widespread nest-box schemes have transformed the fortunes of barn owls across Suffolk and Norfolk, and ringing programmes have proved high survival rates for the nestlings.

What I get from encounters with barn owls is something beyond this. It's partly, of course, their soft, ethereal beauty, especially with the palest birds seen on the cusp of darkness. The plumage on their wings is like marquetry, an intricate mosaic of chequers and lozenges of buff, brown and tan. The patterning on each bird is unique, so that it is possible to tell individual birds apart. Their flight is eloquent and buoyant. Gilbert White remarked that 'they seem to want ballast'. Yet much of the time they scarcely seem like creatures of the air. A high-flying owl looks ridiculous, like an escaped child's balloon. Low-flying birds – and I've seen some hawking in ditches below ground level – are more like emanations of the earth, will-o'-the-wisps. I've watched many owls hunting and find their nonchalant flight ways and paths bewitching. They seem to *lope* much of the time, their gaze fixed on the ground, as the huge sound-reflecting disc of their faces swivel from side to side. They often hover before pouncing. But just as often make a sudden swerve-back, as if they have seen or heard a movement on the ground after they have passed beyond it – the small sound of a vole relaxing as it believes danger has passed, maybe. But I've watched many birds quartering grasslands and failed to spot any orderly pattern in their progress. The same track is repeatedly beaten backwards and forwards, before the bird veers off leaving acres of seemingly similar vegetation unsearched. I am baffled as to why I expect to find an owl's hunting logic comprehensible.

What I think fixes them most powerfully in my heart is the presence they generate, the sense of exactly belonging to the places where you find them. I've found they open up to me a landscape I've not paid proper attention to before: tangled field dykes, grassy glades in stackyards, the offcuts of land left by the hard geometry of modern farming. This is the owls' own parish, an echo of an older landscape which lives inside the matrix of our own.

For them it's a hard-won and precarious belonging. There are years and seasons when they become scarce visions. Their breeding success depends partly on that of voles', which follows a four-year cycle, and in the year after the vole low point barn-owl populations can also crash. In daylight they are mobbed endlessly by crows and are wary of birds of prey. I have seen them strafed by kestrels. Despite the increase in breeding pairs in East

Anglia they are becoming less often seen in the dusk. One theory is that the recolonisation of the region by buzzards, which can frequent the same habitats, has nudged barn owls into leaving the vulnerable and competitive environment of twilight and move into the obscurity of full darkness. They don't like flying in rain, and can't hunt easily in snow or high wind. In this merciless winter they have suffered badly, and emaciated corpses are being found all over the region. Our weather sensitivities have converged, and the owls which at other times console me become sources of empathetic pain. I come back from fruitless searchings with what Polly calls Irritable Owl Syndrome.

But they're ancient British denizens, evolved through our capricious climate, so I go out on my ritual just a few days before Christmas determined not to be pessimistic. I've come to a patch of marshland in a loop of the River Waveney. The temperature is hovering just above zero. The whole landscape is fading to pearl-grey, and grass and icy wallows are indistinguishable. I trudge along ditches of reedgrass blackened by the frost, peer through the opalescent air at tussock glades. I see flashes of white, but they're sleeping farmyard ducks and tumps of unthawed snow. I give up, and am just climbing the fence to the car, when a woodcock rockets out of a ditch beside the fence. I follow it through my binoculars into a wood the other side of the marsh, and glimpse an amorphous lump of white deep in a tree. It is so still and featureless I assume it's a wind-blown plastic bag, a common phantom in the lives of barn-owl seekers. But I feel I should make sure and walk down the lane towards it. Twenty yards away it is still a shapeless bag. Then, as I draw level, it tilts slightly forward. A small bow, whose message to me is quite clear: 'I'm still here, thank you, neighbour, but my existence is not dependent on your observation of me.' Then it spreads its white wings, astonishingly wide for such a hunched creature, drifts off through the black fretwork of the alder trees, and vanishes into a tangle of swamp-patch and river-edge, its own winter retreat.

Acknowledgements

Thanks to the publishers and editors who first commissioned these pieces: Viking Books, Adrian Cooper and Little Toller books, the *Guardian*, Peter Tolhurst and Black Dog Books, *Slightly Foxed*, Kay Dunbar and Green Books, *The Sunday Telegraph*, *The Economist*, *Mondial*, *The New Statesman*, Tim Dee and BBC Radio 3, David Wilkinson and Signal Books, *BBC Wildlife* magazine, Edward Chell, the *Sunday Times*, Jonathan Cape and the *Mail on Sunday*.

For many conversations over the years on the business of memoir and nature writing, I'm indebted to the wisdom and insights of Mark Cocker, Jon Cook, Tim Dee, Richard Holmes, Kathleen Jamie, Robert Macfarlane and James Robertson.

Special thanks to Clara Farmer at Chatto for her patience in waiting so long for this book, and her invaluable comments on its structuring, and to the team at Chatto: Harriet Dobson, Lucie Cuthbertson-Twiggs, Mollie Stewart and Kris Potter.

Thanks also to Clare Roberts for the vignettes at the beginning of each section. They were originally published in *The Journals of Gilbert White* (Century, 1986-9).

Index